THE BRAIN AND PAIN

THE BRAIN AND PAIN

BREAKTHROUGHS IN NEUROSCIENCE

RICHARD AMBRON

Illustrated by

AHMET SINAV

Columbia University Press *New York*

Columbia University Press
Publishers Since 1893
New York Chichester, West Sussex
cup.columbia.edu

Library of Congress Cataloging-in-Publication Data
Names: Ambron, Richard, author.
Title: The brain and pain : breakthroughs in neuroscience /
Richard Ambron; Illustrated by Ahmet Sinav
Description: New York : Columbia University Press, [2022] |
Includes index.
Identifiers: LCCN 2021024031 (print) | LCCN 2021024032 (ebook) |
ISBN 9780231204866 (hardback) | ISBN 9780231204873
(trade paperback) | ISBN 9780231555715 (ebook)
Subjects: LCSH: Pain—Treatment. | Chronic pain—Physiological
aspects. | Chronic pain—Psychological aspects. | Neurosciences.
Classification: LCC RB127 .A437 2022 (print) | LCC RB127 (ebook) |
DDC 616/.0472—dc23
LC record available at https://lccn.loc.gov/2021024031
LC ebook record available at https://lccn.loc.gov/2021024032

Columbia University Press books are printed on permanent
and durable acid-free paper.

Printed and bound by CPI Group (UK) Ltd, Croydon, CR0 4YY

Cover design: Milenda Nan Ok Lee
Cover image: AlexLMX / iStock

CONTENTS

SECTION II
The Modulation of Pain by Circuits in the Brain

THE BRAIN AND PAIN

INTRODUCTION AND NOMENCLATURE

When asked to select our most complex sensation, most will choose vision, hearing, or even smell. That would be incorrect because, as we have learned recently, pain is by far the most complex of all the sensations. It is also the most clinically relevant and, surprisingly, the most important for survival. The purpose of this book is to describe the many recent advances in our understanding of pain and to explain how they are leading to important new approaches to pain management.

To begin, let's consider the fact that our view of the world depends on how accurately the brain interprets incoming information from our senses. In essence, the brain creates a three-dimensional concept of our surroundings by integrating information from vision, olfaction, hearing, touch, pain, and other more subtle sensory systems. This information is communicated to higher centers in the brain, where a response is formulated and a command is sent to our muscles to respond, we hope, appropriately. That all of this happens every millisecond we are awake is quite remarkable. Not every sensation is essential, however, and we can certainly survive without vision, hearing, or smell, and numbness or other disruptions of touch are annoying but tolerable. Pain is the one sensation that is necessary for life because it alerts the brain that an injury has occurred, thereby eliciting a response to protect the wound from additional damage. It is also educational because it teaches us, usually at a young age, what to avoid. People born with defective systems for pain do not survive for long. While protective, pain can be onerous, and we live in an age where pain is viewed as an unwanted intrusion into our lives and is to be avoided. A high percentage of visits to a doctor are prompted by pain, and many hospitals now have entire clinics devoted to pain management. Fortunately, the pain from minor cuts, burns, or abrasions typically diminishes within a day and can be alleviated by over-the-counter medications. Pain becomes a problem when it is prolonged. Post-operative pain, for example, is an issue because it is incapacitating and can last for days or longer. Yet even this type of pain can be managed with powerful analgesics despite the unwanted side effects. Far more serious situations develop when patients suffer from chronic pain, meaning that it lasts for months or even years. Such pain is deemed pathological because it has no benefit and severely diminishes the quality of life. As you can imagine, people who suffer from chronic pain have a difficult time focusing and often suffer from anxiety, fear, and depression. Unremitting pain also disrupts family relationships, and the loss of productivity has negative economic consequences. At any given time there are an estimated thirty million people in the United States who suffer from chronic pain, and for the majority there is no option other than drugs containing opiates; because they are addictive, they have resulted in an epidemic of drug abuse. Data from 2017 attributes seventy thousand deaths to overdoses of opiate painkillers. By any standard, this is a tragedy of epic proportions that can only be remedied by understanding the mechanisms responsible for the pain.

Recent advances in neuroscience have provided many new insights into the neurobiological underpinnings of pain, and we now recognize that pain is not some mystical malady that arises in somehow unknowable ways. In fact, pain in response to a lesion, whether an injury or an inflammation, is mediated by well-defined, stereotypic neural pathways, and we will devote several chapters to discussing the molecular, cellular, and neuroanatomical components of these pathways. This information provides an essential background for the prevailing idea that chronic pain is caused by malfunctioning proteins. Hence, the goal of the pharmaceutical industry is to develop drugs that will attack these rogue molecules and alleviate the pain. This pharmacological approach has merit, but there are several obstacles, primarily the overwhelming complexity of the nervous system and the fact that many of these proteins have functions in other systems so that interfering with their action in one system will always have side effects in another.

This book is divided into sections that view pain from a variety of perspectives. The first section examines the premises that underlie the target-based pharmacological approach and discusses the pathways that are responsible for communicating information about a lesion to centers in the brain. This study will require some knowledge of neuroscience, and we will therefore devote several chapters to neuroanatomy and cellular and molecular neurobiology. Don't be alarmed. Although there are many thousands of articles describing relevant research, our goal in each chapter is to provide sufficient information so that the reader can understand the fundamental processes and issues. This comprehension is made easier because the information in the text is reinforced by diagrams and illustrations. In addition, we will focus only on the molecules and events that we believe are most relevant to understanding the initial responses to pain.

Understanding how information about an injury is conveyed to the brain required many decades of research, but we now know that this is only a relatively small part of the story. The complexity of pain actually arises when the brain receives this information because the degree to which pain is experienced is highly subjective and is influenced by past experience, present circumstances, beliefs, and a variety of other factors. Until recently, we had no inkling as to how these factors could modulate pain. This all changed when advances in real-time imaging enabled clinicians and neuroscientists to visualize the activity within the brain of patients in pain. The images revealed that the intensity of the pain correlates with the activity of discrete clusters or groups of neurons that can be mapped as modules in what we will call a *pain matrix*. Remarkably, these studies indicated that all pain emerges from interactions among these modules, including the anguish that arises from the death of a loved one. Thus, suffering from both physical injury and psychological trauma share circuits in the brain. Needless to say, these findings have dramatically changed our understanding of pain, and we will discuss how they are leading to new approaches to pain management.

New approaches are certainly welcome because contemporary Western medicine has not been very successful in treating pain, especially chronic pain. In fact, most of our most successful analgesics are only refinements of agents, such as opium, that have been used for centuries to relieve pain. We will discuss how opium relieves pain and will also describe promising new developments into the analgesic properties of marijuana. Nevertheless, many forms of alternative medicine have appeared; the popular literature contains a bewildering variety of books that claim to have found novel approaches to reducing pain. Ranging from herbal mixtures to correcting misaligned or conflicting forces within the body, the vast majority of these treatments lack a scientific basis for their claims. For this reason, they have been largely dismissed by most physicians. Of course, the various alternative practitioners then counter this rejection by citing numerous endorsements extolling their success. Even the so-called snake oil salesmen who sold "magical" elixirs could creditably claim to have reduced pain in some of their customers. In fact, many of these alternative approaches have no actual therapeutic benefit but will have some measure of success because of a fascinating phenomenon known as the placebo effect: in which a patient's pain is reduced if they believe that the remedy is real. We now understand the basis of this effect in the brain and we will see how this has had a major influence on how to manage pain.

Eastern societies took a different route to manage pain with the introduction of meditation and its practitioners have claimed for thousands of years that training the mind can attenuate pain. These claims were met with considerable skepticism because they invoked the presence of mysterious energies or forces whose existence cannot be experimentally verified. We can now state that the skepticism is no longer warranted because recent studies have shown that meditation has a firm foundation in neuroscience and that the ability of practitioners to willfully modulate pain provides an important alternative to drugs for the treatment of chronic pain. These possibilities are discussed in several of the later chapters in the book.

NOMENCLATURE

In theory, most science deals with facts that are expressed in words, so the meaning of the words is very important. Consequently, we must be aware that certain words that are understood in common usage have deeper meaning when discussing pain. For example, *pain* itself is a term whose definition depends on perspective. The International Association for the Study of Pain (IASP) defines pain as "an unpleasant sensory and emotional experience associated with actual or potential tissue damage, or described in terms of such damage." While this definition is correct, it is insufficient because it omits properties of pain that are important for understanding its origins. Thus, pain varies in intensity from merely unpleasant to unbearable and also has texture—it can be sharp, dull, or burning. In addition, pain as a response to an injury provides awareness of the injury but becomes "painful" depending on the activity of dedicated circuits in the brain. We cannot understand pain unless we can explain how these properties arise from actions within the nervous system.

The word *sense* typically means an awareness of a stimulus, be it touch, itch, etc. From a scientific standpoint, however, a sense is a *perception* of a specific type of stimulus that emerges from specific circuits in the brain and is inextricably bound to consciousness. This relationship has important implications for pain that will be discussed in detail. Likewise *injury* is understood to be a consequence of an event that has caused pain. Such events are considered *noxious.* Synonyms include *lesion* and *insult.* Injury also has a connation of physical damage, such as a cut through the skin or a torn muscle or ligament, but we will learn that pain can exist when there is no obvious physical tissue damage. The most dramatic example is the anguish due to grieving, but certain types of inflammation can also elicit a pain without an obvious source.

We must also be careful when using the term *chronic pain,* which is defined differently depending on one's perspective or field of medicine. For our purposes, chronic pain is defined by two features: it lasts longer than three months and it is associated with changes in gene expression in the neurons that comprise the pain pathways. We will further refine this definition as we advance our knowledge of the nervous system.

All sciences have developed a vocabulary that enables practitioners to communicate, but which can be unintelligible to those outside the field of specialty. This vastly complicates the writing of a book on pain because to cover the topic adequately requires a vast terminology that encompasses not only traditional human anatomy but also neuroanatomy, cellular and molecular neurobiology, and biochemistry. Each discipline has its own vocabulary, and we cannot escape using terminology that is unfamiliar to most readers. This is especially true when discussing all of the proteins, peptides, and other agents that are essential contributors to pain. To make matters worse, the nomenclature is confusing because in many cases, these compounds were named at the time they were discovered, but scientists learned at a later date that their function is quite different. To avoid overwhelming the reader, we will limit ourselves to discussing only those molecules that are either important mediators of pain or potential targets for the development of drugs to manage pain. By way of encouragement, we can state unequivocally that merely knowing that such molecules exist, without necessarily remembering their names, is all that is needed to understand how noxious events give rise to pain. Finally, scientific journals use acronyms as a way of reducing space. For example, adenosine triphosphate is commonly known as ATP. However, acronyms can be annoying because they often force the reader to go back to find the original word. We will use acronyms, but we will repeat the full name when the terms are separated by one or more pages.

I

THE BASIC PAIN PATHWAY AND THE MOLECULAR MECHANISMS THAT DETERMINE THE INTENSITY AND DURATION OF PAIN

1

PAIN AS A PROPERTY OF THE NERVOUS SYSTEM

OVERVIEW: PAIN IS INSTRUCTIVE, NECESSARY, AND ADAPTIVE

Before we begin to discuss the nervous system in detail, a few concepts are worth mentioning. We are aware of vision, hearing, smell, and touch because they are continually experienced. Pain is different because it is not present under normal circumstances and is usually transient. It is a complex sensation that emerges only when information from the site of an injury or inflammation reaches processing centers in the brain. This information is conveyed by an extensive network of nerves, but the pain does not exist at the site of the lesion or as a component of these nerves. Rather, pain is *perceived* as a sensation only when the information activates circuits in the brain. This is an important concept to grasp and is analogous to turning on a light bulb: throwing the switch (the injury) generates a current (a signal) that travels through a wire (a nerve) that activates the bulb (the brain). Although pain is considered onerous, it is essential to life. It alerts us that our body has been damaged and motivates us to protect the site until the lesion has healed. It is also a powerful teacher. During childhood we learn much about our environment, especially about things that can injure us, such as touching a hot stove or the edge of a knife. Because injuries hurt, we learn to avoid situations that can lead to an injury. Consequently, people who can't perceive pain usually don't survive.

Now, let's suppose you sustain a minor cut to your hand. The immediate response is a rapid withdrawal of your injured hand to prevent more damage. This action is followed by the onset of a sharp or *acute pain*, which actually follows milliseconds after pulling your hand away. The delay makes sense because the withdrawal can occur much more quickly than the finite amount of time it takes for a signal from the injury site to reach the brain and be interpreted. The acute pain makes us aware of the injury and that the degree of pain is commensurate with the seriousness of the injury. Acute pain diminishes rapidly if the injury is minor but will undergo a transition to persistent pain if the injury is more serious. This keeps us aware of the injury so we continue to protect the injured hand. Persistent pain disappears once the injury has healed. Hence, the response to pain is not fixed but is *adaptive*—both the intensity and duration of the pain vary with the severity of the injury. We know this from our own experience: the more severe the injury, the more intense and longer lasting the pain. Adaptation is an inherent property of the pain pathways and is therefore governed by specific events within these pathways. Put another way, the pathways to the brain are constant, but the information that is communicated via these pathways is malleable and can be modified by circumstances. Chronic pain, on the other hand, is a pathological condition that serves no protective purpose and is not adaptive. We can view chronic pain as an abnormally prolonged version of persistent pain. From this perspective, both normal and pathological pain are governed by events inherent to the nervous system pathways that are responsible for the pain. In chronic pain conditions, however, something has gone awry in one or more of the key events involved. Consequently, a guiding principle in treating chronic pain is to identify these key events and develop drugs that block them and alleviate the pain. This approach appears logical, but as we will see, it has proven to be extraordinarily difficult to implement.

In addition to these inherent processes, the response to pain is further complicated by the fact that the intensity of the pain is subjective and can be modified by the context in which it is experienced. Thus, an injury that would be extremely painful in one setting will be much less so if it occurs in a situation where your life is in danger. For example, you are walking in the woods and twist your ankle. You sit down on a log in considerable pain, but if a bear were to suddenly appear, you would get up and run as if you weren't in pain. Aside from the fact that you should not run from a bear, you are experiencing what is known as *stress-induced analgesia*, i.e., a painful injury that would be incapacitating under normal circumstances can be ignored in order to escape death. On the other hand, dwelling on pain only makes it worse. Anticipation generates anxiety that can also increase pain, as when a nurse approaches with a needle to give an injection. Stress-induced analgesia and exacerbation of pain by anxiety are not direct properties of the pain pathways but are imposed on these pathways by circuits in the brain. Until recently we knew very little about these circuits, but recent advances in neuroscience have greatly expanded our understanding of how they function and their importance in managing pain.

SENSATIONS AND THE NOTION OF SELF

We also need to have a general understanding of the relationship between the brain and the processes that mediate sensations.

When we talk about the notion of "self," we typically envision our body, a corporeal entity comprised of a heart, lungs, brain, digestive organs, etc. However, from another perspective, the notion of self is actually a manifestation of consciousness, which is a property that emerges uniquely from the circuitry in the brain.[1] Recognizing this duality is significant because the brain lies ensconced in the skull. Consequently, the only way that the brain can be made aware of the external world is via its connection to the nerves that convey signals from the skin, eyes, ears, nose, and tongue to the brain, where they are then translated into sensations. Pain is one of these sensations, but remember that the perception of a sensation also arises from circuits in the brain, meaning that sensations and consciousness are inextricably linked. It is the integration of all these sensations that makes us aware of our surroundings and, by connecting to motor systems, allows us to manipulate our environment.

In addition, we need to recognize that as far as the brain is concerned, there are actually two external worlds. The first is the world that surrounds us; the second is the internal world of our organs. Just as it is clearly important to be informed of lesions to the skin, it is also important that the brain be made aware of threats to the function of the heart and other organs since they maintain the supply of nutrients and oxygen to the brain. Pain is the primary response to a lesion in the viscera, usually a distension, such as a calculus (stone) in the ureter, or an inflammation. Thus, pain informs the brain of threats from both the external and internal worlds.

Given those distinctions, we will take a step-by-step approach to understanding how we experience pain. We will first form a foundation by discussing neurons and how the formation of neuronal networks provides a way to communicate information rapidly and over long distances. We will then move on to the organization of the human nervous system and the neuronal pathways that convey the signals for pain. We will use examples from everyday experience to anchor the text to reality. Next, we will focus on how molecular changes inherent to these pathways alter the perception and duration of pain. Emphasis will be placed on those changes that provide the most promising targets for the development of analgesics. We will then describe how complex networks within the brain modulate the experience of pain. These networks are very important in understanding more recent, non-pharmacological approaches to alleviating chronic pain.

NEURONS, PRIMITIVE NETWORKS, AND REFLEXES

The survival of any animal species depends on its ability to respond to threats from the environment; even single-celled animals are capable of withdrawing from such threats. As evolution progressed to multicellular animals, however, the ability to react to threats became more complicated. Take, for example, the freshwater polyp *Hydra vulgaris*, whose body is a partly hollow cylinder with a solid base and an opening at the mouth end (see fig. 1.1A). The mouth leads into a hollow interior that acts as the digestive cavity and is surrounded by tentacles that facilitate the acquisition of food. A layer of ectodermal cells lines the surface and a layer of endodermal cells lines the internal digestive area (see fig. 1.1A, inset).

Note that all these cells are exposed to the outer world, a relationship to keep in mind because it is also present in humans, although in a much more complex form. Between the layers are muscle cells that regulate movements. In such multicellular animals, the cells on both internal and external surfaces cannot

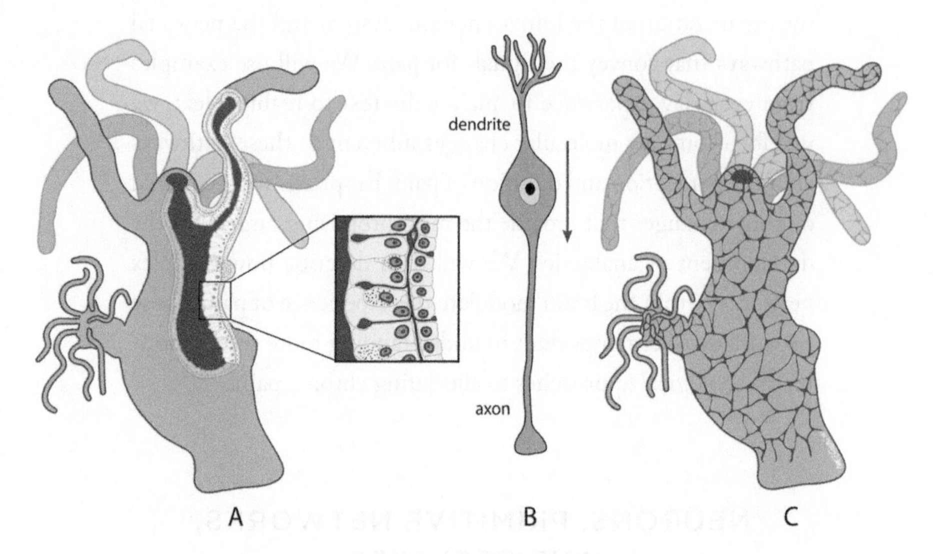

FIGURE 1.1 The *Hydra vulgaris* nervous system. (A) Hydra with an opening to expose the internal cavity (black) and body wall. Inset: an enlarged segment of the body wall showing both the internal and external layer of cells and the interspersed neurons with their receptive extensions (dendrites) exposed on the external and internal surfaces. (B) A neuron with its cell body, dendrite, and axon. (C) Intact hydra showing that the neurons and their extensions form a neural network.

respond individually to a threat since the movements of the body are effected by the muscles and must be coordinated. The solution was the appearance of specialized cells called neurons (or nerve cells) that differ from typical round or rectangular cells in having two processes that extend from the cell body and can project long distances (see fig. 1.1B).[2] A short process known as a *dendrite* reaches the surface of the animal, where its terminal membrane contains receptors that are exposed to the surroundings and can respond to external cues. The longer *axon* process courses within the animal, where it can terminate on other neurons, forming a network, or on muscle cells.[3] The network is dynamic because the processes are electrically excitable and can rapidly conduct an electrical signal known as an *action potential*, much like current passing along a wire. When the receptors are activated, an action potential is generated that propagates rapidly along the neuronal network to the muscle cells, which causes them to contract. This results in a coordinated *reflexive withdrawal* of the tentacles and body that greatly reduces the animal's size and exposure. In a similar fashion, the neurons whose processes are exposed to the digestive cavity can respond to agents that have entered or been captured by the tentacles. Thus, in this simple organism we see how neurons can form simple networks that allow information to flow rapidly from internal and external surfaces to cells within the animal that can respond (fig. 1.1C). The appearance of the neuron is therefore one of the most prominent events in animal evolution. In higher animals and humans, neurons mediate reflexive activities but also carry out more refined responses, including the transmission of information that enables us to perform complex tasks and to inform us of a harmful injury or lesion via the sensation perceived as pain.

2

ORGANIZATION OF THE HUMAN NERVOUS SYSTEM

From Nerves to Neurons

PERIPHERAL NERVES AND DERMATOMES

If we accept the premise for now that chronic pain arises due to a malfunction in some component of the nerves responsible for pain, then we should begin by discussing the general organization of our nervous system. Anatomists traditionally consider this system to have two parts; a central nervous system (CNS) consisting of the brain and its continuation, the spinal cord, and the peripheral nervous system (PNS) consisting of the nerves that emerge from the CNS and course throughout the body. Because the brain is the most complex structure known, it was therefore far beyond the capabilities of early anatomists to try to understand how it functions.[1] In addition, it is ensconced in the skull, which limits access, and attempts to manipulate the brain typically resulted in disastrous disruptions in function. The spinal cord is nowhere near as complex, but it is enclosed in a canal formed by the bones and ligaments of the vertebral column and is similarly susceptible to damage. In contrast, the nerves in the PNS were able to be examined by physicians after serious injuries and were studied by anatomists for centuries. We have therefore acquired a great deal of knowledge about their distribution and function.

The number and origin of the peripheral nerves are constant from individual to individual, and they are classified by where they emerge from the spinal cord (fig. 2.1).[2] Twelve pairs of cranial nerves emerge from the brain,[3] and thirty-one pairs of spinal nerves arise sequentially from the spinal cord (fig. 2.1A). The vertebrate nervous system is bilaterally symmetrical so that one nerve of each pair supplies the right side of the body and the other the left side. We will focus on the spinal nerves for now and will briefly describe how their distribution is important for an understanding of pain.

Each spinal nerve is formed just outside the vertebral column by the merger of a dorsal (posterior) and ventral (anterior) root (fig. 2.1B). Residing on the dorsal root is a swelling known as a *dorsal root ganglion* (DRG) that will become important when we discuss its contents. After a short distance, each spinal nerve divides into a ventral (anterior) ramus, or branch, that courses to the front of the body and a dorsal (posterior) ramus that extends to the back. The distribution of the dorsal and ventral primary rami is easy to understand in the thoracic region, where they emerge in sequence and circumnavigate the body in a groove beneath the ribs (fig. 2.1C). Thus, we can number the thoracic nerves by their origin from the spinal cord (T1, T2, etc.). In contrast, the nerves that emerge in the cervical, lumbar, and sacral regions intermingle to form a plexus (fig. 2.1A), which reflects the development of the upper and lower limbs where the skin and underlying structures are not neatly ordered, as in the thorax. Unlike the thoracic nerves that originate from only one spinal cord level, those that emerge from each plexus contain components from several spinal cord levels and are given names, such as the median, ulnar, etc.

Emerging from each ramus along its course are many small branches that terminate on a defined band of skin known as a

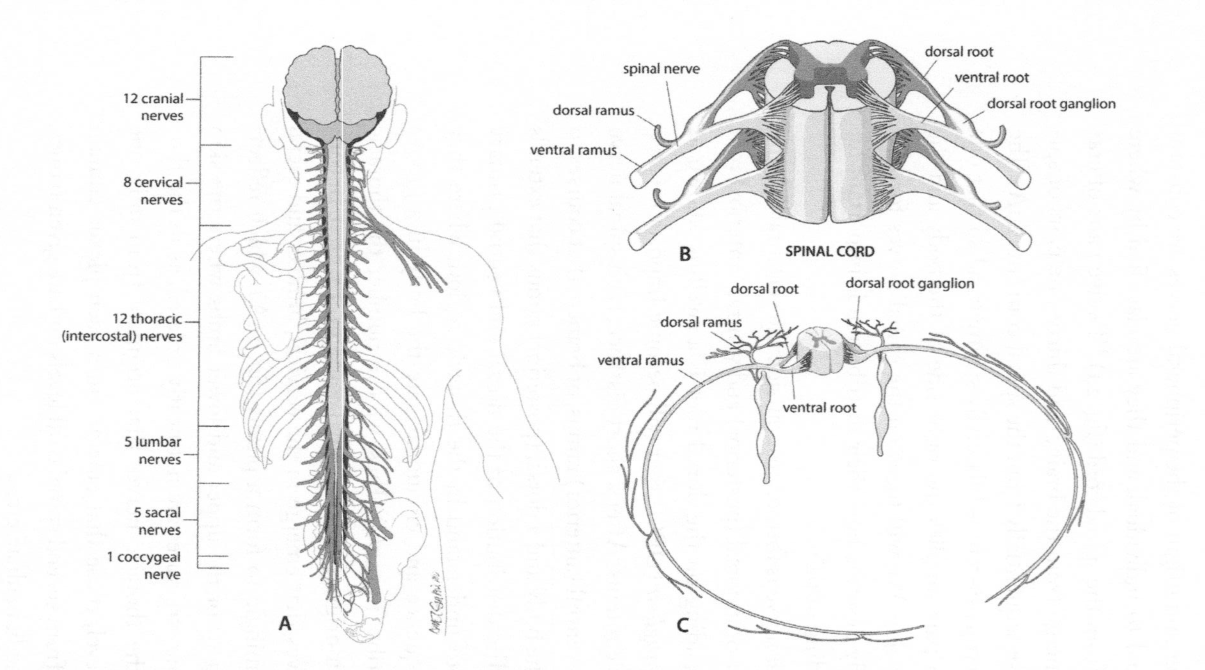

FIGURE 2.1 The origin and distribution of the nerves comprising the peripheral nervous system. (A) Twelve pairs of cranial nerves emerge from the cranial region and thirty-one pairs of spinal nerves emerge from the spinal cord. The twelve thoracic nerves emerge in sequence and on the left side are shown as they course just beneath each rib. On the right side, each of the nerves in the cervical, lumbar, and sacral regions are shown intermingling to form a plexus. (B) The formation of a spinal nerve. Small branches that emerge from the dorsal and ventral region of the spinal cord form a dorsal root and a ventral root, respectively. A dorsal root ganglion sits on the dorsal root. The two roots merge to form a short spinal nerve that divides into a small dorsal ramus (branch) to the back and a larger ventral ramus (branch) to the front. (C) A single thoracic nerve with its small dorsal branch and large ventral branch that encircles the body. Small branches from each innervate muscles and other structures and terminate in the skin.

dermatome, as well as on the muscles, bone, blood vessels, and sweat glands that lie beneath each dermatome (fig. 2.2). Thus, each dorsal ramus supplies the contents of a specific dermatome in the back and the ventral ramus the portion in the front.[4]

Although there are small individual variations in the dermatome pattern, the path of each nerve branch as it passes throughout the body to its dermatome is reasonably constant. For example, an injury just to the left of the umbilicus (belly button) will involve the tenth thoracic dermatome that is innervated by branches of the tenth thoracic spinal nerve that will

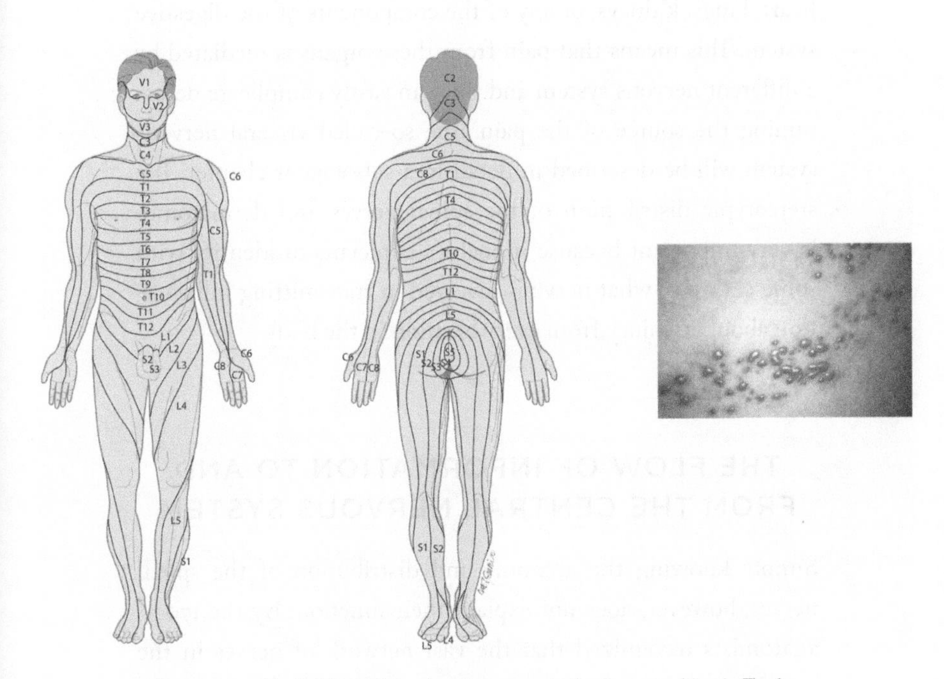

FIGURE 2.2 (*Left*) Map of the dermatomes on the front and back. Each dermatome is innervated by the dorsal and ventral branches of a specific spinal nerve. Thus, the region of the umbilicus (belly button) is supplied by the anterior (ventral) branches of thoracic spinal nerve T10 that emerges from the tenth thoracic level of the spinal cord. (*Right*) Blisters arising from *Varicella zoster* (shingles) along a thoracic dermatome on the side of the body.

communicate with the left half of the spinal cord at the tenth thoracic level. The path of the nerves in the thoracic region is very obvious in the condition known as shingles, which arises in adults from a reactivation of the *Varicella zoster* virus that causes chicken pox (see fig. 2.2). Shingles can affect any peripheral nerve throughout the body and manifests as very painful blisters in the skin of the dermatome supplied by the infected nerve.

Another name for body is *soma* and the peripheral nerves are therefore also known as *somatic nerves*. Most notably, the soma does not include our viscera, i.e., the internal organs such as the heart, lungs, kidneys, or any of the components of the digestive system. This means that pain from these organs is mediated by a different nervous system and this can vastly complicate determining the source of the pain. The so-called visceral nervous system will be described in detail in a subsequent chapter. The stereotypic distribution of the spinal nerves and dermatomes is very important because it enables clinicians to identify with some certainty what nerve is involved in transmitting information about an injury from every location in the body.

THE FLOW OF INFORMATION TO AND FROM THE CENTRAL NERVOUS SYSTEM

Simply knowing the anatomy and distribution of the spinal nerves, however, does not explain their function. By the 1700s, anatomists recognized that the vast network of nerves in the peripheral nervous system was responsible for detecting the stimuli responsible for touch, pain, and temperature from the skin and was sending this information to the mysterious CNS.[5] Note that we use the word "detect" and not "sense." This distinction is important because the ability to sense—to be aware of a detected stimulus—depends on the activation of highly sophisticated circuits in the brain that will be discussed later. Early anatomists termed the information flowing toward the CNS as *afferent* and determined that the vast majority of this information reaches the spinal cord via the dorsal roots. Likewise, they knew that if a nerve was cut, the muscles to which it was connected became paralyzed. This meant that information from the CNS was flowing outward along the nerves, and it was termed *efferent*. All efferent information exits the CNS to the spinal nerves through the ventral root.[6] Thus, it is the transmission of afferent and efferent information via the nerves in the periphery that enables us to manipulate our surroundings. For example, when we attempt to thread a needle, afferent signals from the visual system and tactile information from our fingers travel via nerves to the brain where they result in our seeing the opening and feeling the needle and the thread. Since our motivation is to thread the needle, the appropriate neuronal circuits in the brain are activated and efferent information flows outward along the nerves to the muscles that enable us to push the thread through the opening. What these early anatomists could not discern were the structures actually responsible for conveying the afferent and efferent information along the nerves. They only became apparent after the development of the microscope.

MICROSCOPIC NEUROANATOMY: THE NEURONAL BASIS OF PAIN

The invention of the microscope, attributed to the Dutch scientist Antonie van Leeuwenhoek in the mid-1600s, was one of history's most significant scientific developments because it opened worlds that could not be resolved by vision alone. For the first time, scientists could study the cells that form the basic units of life. This was facilitated by the application of specific dyes and other procedures that made it possible to distinguish one cell type from another. Special staining techniques developed by Santiago Ramón y Cajal were invaluable in this regard.[7] When scientists trained their microscopes on the nervous system that had been stained, they entered the hitherto unknown realm of the neuron. Indeed, when they looked inside a spinal nerve they saw thousands of fibers (fig. 2.3). Thus, the nerves that were described by ancient anatomists and studied by contemporary anatomists were merely conduits; it is the fibers within that actually convey the information. By tracing these fibers it was soon established that they are extensions of neuronal cell bodies that are located a great distance away.

We first encountered simple neurons in the *Hydra vulgaris*, but the neurons in the human body come in many sizes and shapes

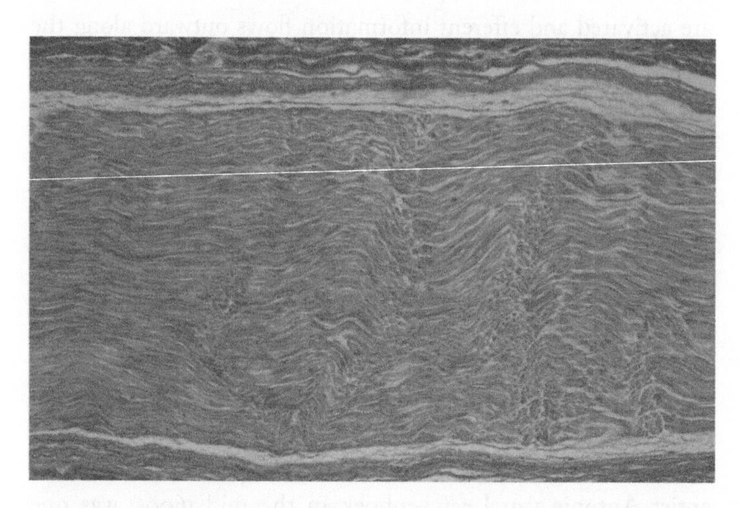

FIGURE 2.3 Microscopic view of a section through a stained peripheral nerve showing that it is composed of thousands of nerve fibers coursing to and from the periphery.

that reflect the need for much more complex functions. For our purposes we will focus on the two types that predominate in the PNS (fig. 2.4). The first of these, the *motor neuron*, has a cell body that is located within the spinal cord at each of the thirty-one levels. The cell body contains the genomic material and the apparatus that manufactures the proteins and other macromolecules needed to maintain the various neuronal functions. Extending from the cell body are several short dendrites, and a very long single axon. The axon exits the spinal cord via the ventral root to enter the appropriate spinal nerve. It then courses within either the dorsal or ventral ramus of the nerve to its target muscle in the dermatome. Like the neurons in *H. vulgaris*, neurons in the human nervous system also respond to a stimulus by generating an electrical impulse—an action potential—that can propagate rapidly along dendrites and axons. We will have much more to say about the generation of action potentials in a subsequent chapter. For now we should understand that a stimulus to the dendrites of a motor neuron will elicit action potentials that will travel along its axon and result in the contraction of its target muscle. Motor neurons are the efferent arm of the PNS and are responsible for all muscle movements.

All afferent information to the central nervous system is conveyed by neurons whose morphology is very different from that of the motor neurons (fig. 2.4). The cell body of all *afferent neurons* resides in the dorsal root ganglion, which, as we mentioned earlier, sits on the dorsal root. A single short projection exits the cell body and divides into two processes. The much longer *peripheral process* enters a spinal nerve and is distributed to its dermatome via either the dorsal or ventral branch. The shorter *central process* enters the dorsal root and courses into the spinal cord where it communicates with spinal neurons. Like motor neurons, afferent neurons are electrically excitable and action

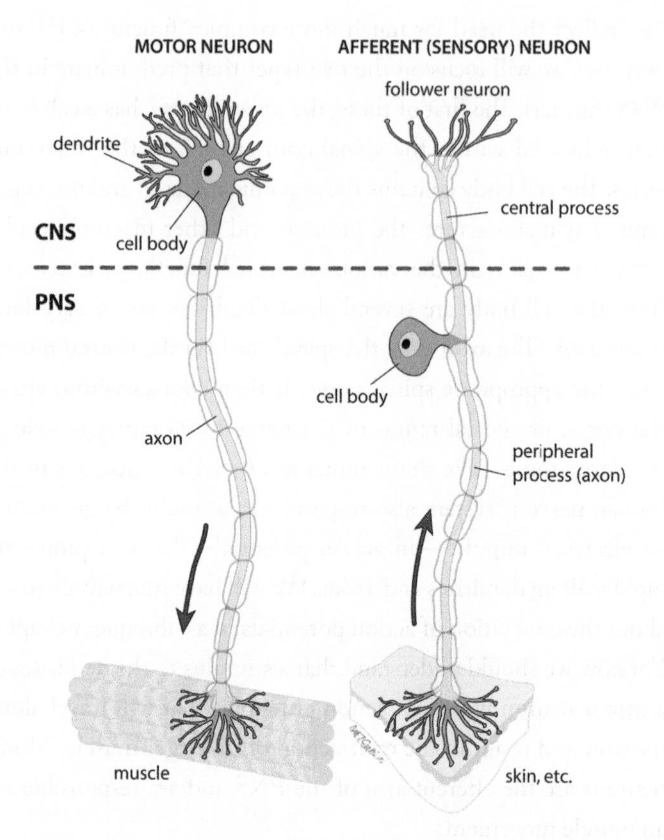

FIGURE 2.4 Morphology of motor and afferent neurons. The long axon of the motor neuron innervates muscle cells, and the long peripheral process of the afferent neuron conveys sensory information from the skin and other structures in the periphery. Both are located within peripheral nerves. The cell body of the motor neuron and the termination of the short central process of the afferent neuron are located within the central nervous system. The cell body of all afferent neurons resides in dorsal root ganglia.

potentials elicited in the peripheral axon will continue to and along the central process. Thus, it is via action potentials that information from structures in the periphery, such as our skin, is conveyed to neurons in the spinal cord.

Using the term processes (or fibers) to describe the peripheral branch of afferent neurons is correct but cumbersome. According to traditional neuroanatomy, axons conduct action potentials away from the cell body. However, the peripheral process of afferent neurons clearly conveys information toward the cell body, violating this convention. More recently axons have been defined not by the direction of conduction but by their limited ability to synthesize proteins. By this criterion, the peripheral process can be called an axon. We have opted to use both axon and process.

Given their architecture and electrical properties, it is easy to see why these neurons are classified as afferent. Note that the cell bodies of the motor neurons are located in the CNS, as are the terminations of the central processes of the afferent neurons. Hence, dividing the nervous system into a CNS and PNS is purely an anatomical construct. Both motor and sensory functions have no such boundary and are completely integrated. Each of the thirty-one pairs of spinal nerves contains the axons of thousands of motor neurons and the peripheral processes of thousands of afferent neurons that will communicate with structures in the periphery. These were the wavy, stained structures seen in figure 2.3. To give some idea of scale, the motor axons and peripheral afferent axons that innervate our big toe can be over three feet long, yet each is thinner than a hair!

The realization that afferent neurons are somehow responsible for communicating sensations was a major advance, but the question that still bedeviled anatomists was how sensory neurons distinguish between touch, pain, temperature, and other stimuli? The answer is that the quality of a sensation (also known as a modality) does not depend on the stimulus but on the properties of the neurons that respond to the stimulus. In other words, there are sensory neurons that respond to touch, others to pain, and so on for each modality. Recent evidence indicates that there is even a subset that responds to itch. The cell bodies of all types are found in the dorsal root ganglia and their peripheral axons enter all of the branches of the spinal nerves associated with that ganglion. We are concerned only with those that respond to events that will result in pain; these are known as *nociceptive neurons* (or *nociceptors*). Charles Sherrington (1857–1952) first introduced the term *nociception*. *Noci* is Latin for pain and these neurons are the first responders to a lesion and are therefore known as *first-order nociceptive neurons*. They differ from the neurons that mediate our other senses because they respond only when there is an injury or other lesion and are therefore silent for much of the time.

NOCICEPTIVE NEURONS AND THE BASIC PAIN PATHWAY

Work by neuroscientists in many disciplines has enabled us to describe in some detail the microscopic anatomy and function of the first-order nociceptive neurons.[8] Thus, we are now ready to place the nociceptive neurons in their proper place in the peripheral nervous system and will use a practical example. An injury has damaged the skin in dermatome T10 near the umbilicus (fig. 2.5). By a mechanism that we will explain later, the injury will elicit action potentials at the terminal of the peripheral process of the first-order neuron. The action potentials will then rapidly propagate along the peripheral process of the neuron within the ventral branch of nerve T10. They will continue to the central process in the T10 dorsal root to the spinal cord where it divides. Each branch communicates with a unique target. One will ultimately result in the activation of a motor neuron whose axon will exit through the ventral root and course within the ventral

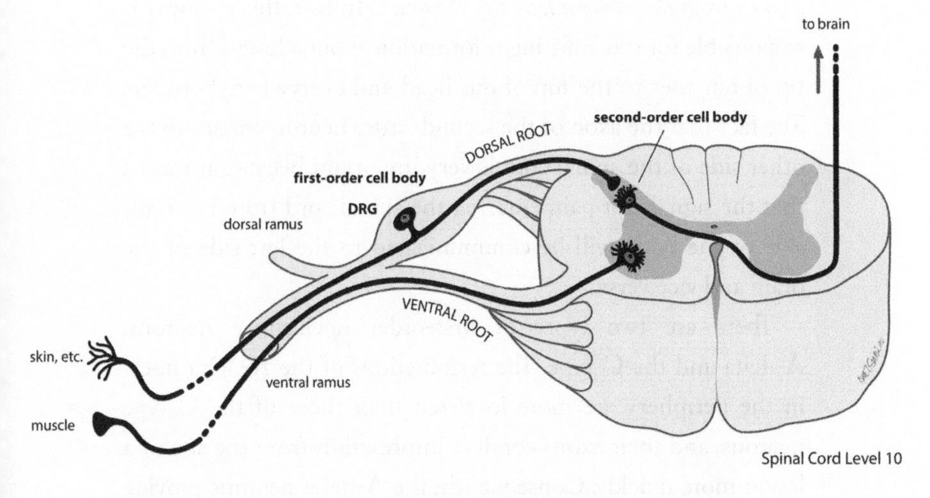

FIGURE 2.5 Schematic showing the response to an injury. Action potentials elicited by an injury to the skin propagate along the peripheral and central process of first-order C-type neurons to activate second-order neurons in the dorsal region of the spinal cord and (indirectly) motor neurons in the ventral region. Action potentials elicited in the second-order neurons course along their axons that cross to the opposite side and ascend to the brain. Action potentials elicited in motor neurons propagate along their axons that exit via the ventral root and course within a branch of a spinal nerve to evoke contractions of target muscles, thereby protecting the injured area from additional damage.

branch of T10 to cause a contraction of muscles at the injury site. We can now understand the rapid reflexive withdrawal that protects the site from more damage. The other branch is even more significant because it elicits action potentials in a *second-order nociceptive neuron*. These potentials will course within the axon of the second-order neuron as it crosses to the other side of the spinal cord and ascends to the brain (see fig. 2.5). What we have just described is quite remarkable: *this basic first order-second order, two-neuron pathway is responsible for communicating signals from the site of a lesion to the brain.* In fact, this pathway is responsible for transmitting information about a lesion from the tip of our toes to the top of our head and everywhere between. The fact that the axon of the second-order neuron crosses to the other side of the spinal cord is very important because it means that the signals for pain entering the spinal cord from the right side of the body will be communicated to the left side of the brain and vice versa.

There are two types of first-order nociceptive neurons: A-delta and the C-type. The terminations of the A-delta fibers in the periphery are more localized than those of the C-type neurons, and their axons conduct information from the site of a lesion more quickly. Consequently, the A-delta neurons provide the almost instantaneous response to an injury. However, there is overwhelming evidence from clinical and experimental studies that C-type neurons are essential for understanding the pain that follows, especially severe persistent pain. We will therefore focus on these first nociceptive neurons in the following chapters.

The two-neuron nociceptive pathway just described explains how signals from an injury to the structures of the body are communicated to the brain. But remember that the term "body" does not include the viscera. How the brain is informed about a lesion in the heart or stomach, for example, is important because several types of chronic pain are associated with our internal organs. The pathway involved is complex, and we will defer a discussion of its function to a subsequent chapter.

3

PAIN

Perception and Attribution

Figure 3.1 depicts the *basic nociceptive pathway* in more detail. In particular, it shows an interneuron imposed between the central processes of the first-order nociceptive neurons and the motor neurons. Much more important for our understanding of pain, however, is that the other branch of the central processes does not directly contact the second-order nociceptive neurons: there is a gap in between that is part of a specialized structure known as a *synapse*. The synapse between the first- and second-order neurons is especially significant in controlling the access of pain signals to the brain, and we will discuss the morphology and function of this synapse in the next chapter.

Up to this point, the information about an injury or lesion exists solely as action potentials that are coursing along the basic nociceptive pathway. What happens next is that circuits in the brain will convert these impulses into the sensation of pain and will also recognize the location of the lesion. How this occurs is both amazing and mysterious and can only be appreciated by first learning a bit about the structure of the brain.

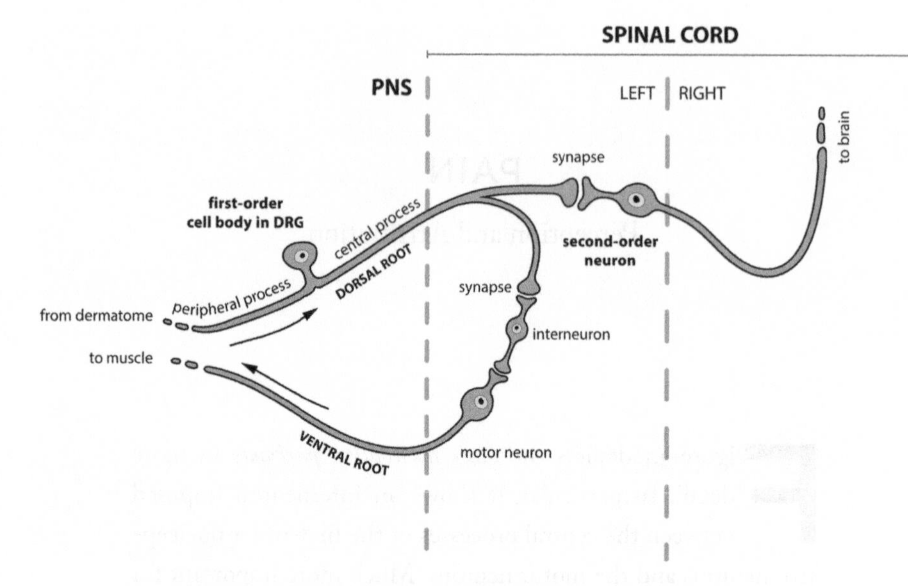

FIGURE 3.1 Schema detailing the synaptic connections formed in the dorsal region of the spinal cord between the central process of a first-order C-type nociceptive neuron and the dendrites of both a second-order nociceptive neuron and an interneuron that communicates with a motor neuron. PNS, peripheral nervous system.

THE BASIC ANATOMY OF THE BRAIN

For a variety of reasons, many early anatomists and philosophers viewed the brain as being of little importance. Imagine how surprised they would be to learn that this innocuous appearing structure is an overwhelmingly complex organ with astounding capabilities. We now know that the brain contains approximately one hundred billion neurons, some of which can communicate with another ten thousand other neurons. In total, the brain is estimated to have a trillion neuronal circuits. It is therefore not an exaggeration to say that the brain is the most complex structure in the known universe.

The brain is dominated by the right and left cerebral hemispheres, which together comprise the *cerebrum* (fig. 3.2A). We know from chapter 2 that the axons of the second-order neurons will transmit signals from an injury on the right side of the body

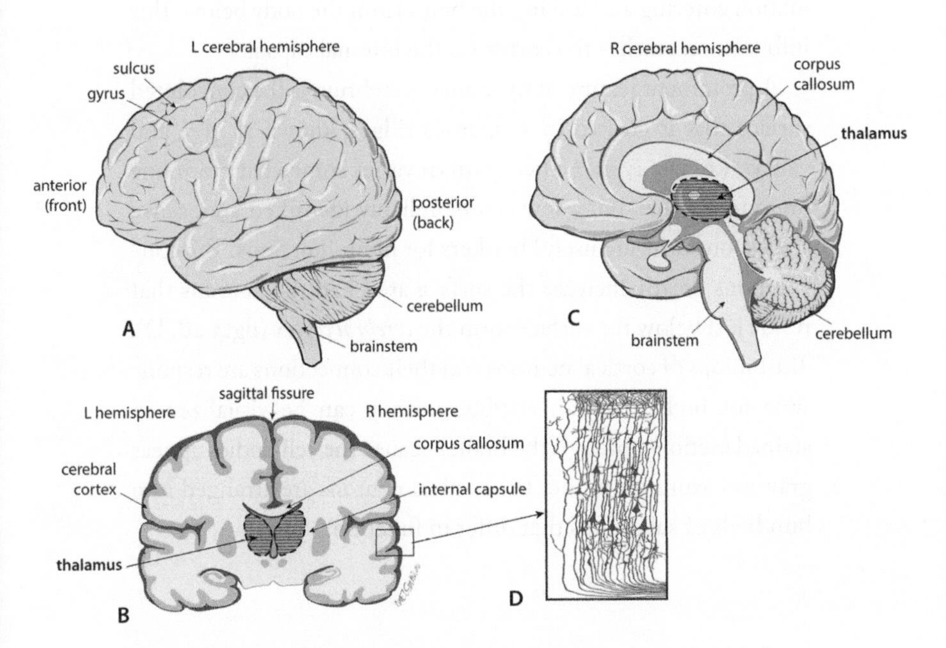

FIGURE 3.2 The human brain. (A) View of the brain from the left side showing the left cerebral hemisphere with ridges (gyri) and valleys (sulci). Below the cerebrum is the cerebellum and the brainstem, which is directly continuous with the spinal cord at the entrance to the skull. (B) Stained cross-section through the hemispheres showing the location of the neuronal cell bodies (gray) in the thin cortex, thalamus (stippled) and axons (white) that course between the hemispheres in the corpus callosum. The two hemispheres are separated by the sagittal fissure. (C) Medial surface of the right hemisphere highlighting the location of the thalamus (stippled). (D) Magnified stained section though the cerebral cortex stained by the method of Cajal showing the arrangement of the cortical neurons and their axons and dendrites that lie just beneath the surface. The billions of cortical neurons form circuits that are responsible for the function of the cerebrum.

to the left hemisphere and vice versa. The hemispheres do not function independently but communicate with each other via a broad band of axons called the *corpus callosum* (fig. 3.2B). Below the hemispheres in the midline is the brainstem, which is continuous with the spinal cord and provides passage for all information entering and leaving the brain from the body below. This information reaches the cortex via the internal capsule.

A prominent feature of the human cerebrum is the convoluted surface that is created by numerous ridges, known as *gyri*, and valleys, known as *sulci*. Their pattern varies somewhat from one brain to another, but a few can be reliably identified and, as we shall soon learn, are useful markers for brain functions. The convolutions greatly increase the surface area, and the neurons that reside just below the surface form the *cerebral cortex* (fig. 3.2B, D). The billions of cortical neurons and their connections are responsible for higher human attributes. They can be visualized in stained sections through the brain because the cell bodies appear gray and axons are white. The cortical neurons are arranged into hundreds of subgroups that differ in function.

PERCEPTION: THE THALAMUS

A key structure for understanding pain is the *thalamus*, which is a paired collection of neuronal cell bodies that resides deep within each hemisphere (fig. 3.2B). Each thalamus is a communication and integration center that receives inputs from all of the sensory neurons in the periphery except those for olfaction, which is a more primitive modality.[1] The thalamus is subdivided into distinct groups of neurons; each of which responds to a specific sensory input (fig. 3.3). Thus, at every instant, nerves conveying action potentials for vision, hearing, touch, and pain enter

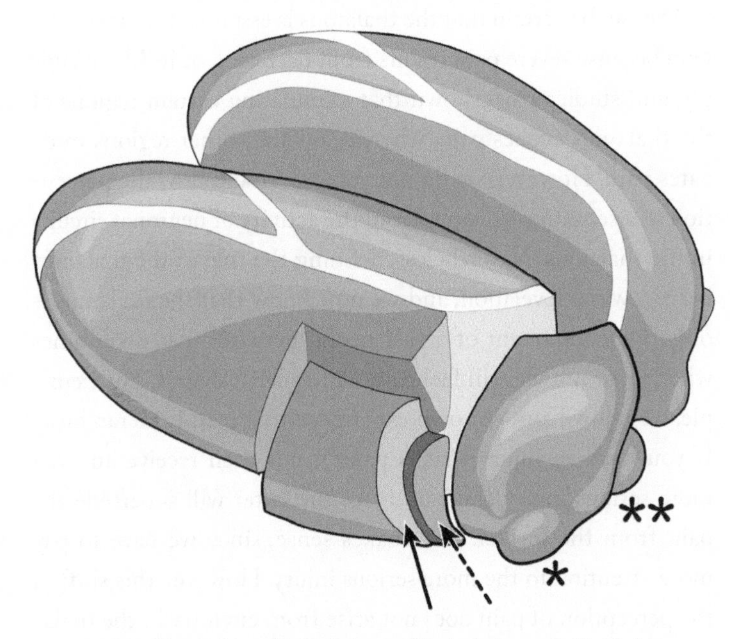

FIGURE 3.3 The right and left thalamus. Each thalamus is divided into regions containing neurons with distinct functions. The thalamus on the left highlights the location of the third-order thalamic neurons that process nociceptive input from the body (solid arrow) and orofacial region (dashed arrow) as well as inputs from the visual (*) and auditory (**) systems. After processing by circuits within each region, information is communicated to centers in the brain.

the thalamus to activate their respective *third-order thalamic* neurons. This information is then relayed independently to other parts of the brain and the outcome is the perception of vision, sound, touch, pain, and all other modalities that we use to create a concept of the world around us. Thus, pain does not exist at the site of the lesion, or in the first- or second-order neurons; *the perception of pain emerges only when the axons of the second-order neurons activate their third-order neurons in the thalamus.* This is a rather profound realization.[2]

We can be certain that the thalamus is essential for perceiving pain because severe pain results from occlusion of its blood supply, and studies have shown that stimulating certain regions of the thalamus evokes pain, whereas ablating other regions mitigates pain. However, we do not yet understand how the perception of a sensation emerges from the activity of neuronal circuits in the thalamus. Nevertheless, defining the role of the thalamus in pain was a revelation, and we now know that the thalamus is only one component of a vast neural network that determines whether or not we will feel pain and to what degree. For example, we know from our own experience that pain is hierarchical. If you have an injury that is painful but then receive an even more serious injury, the pain from the latter will supersede the pain from the former. This makes sense, since we have to pay more attention to the more serious injury. However, this shift in the perception of pain does not arise from circuitry in the thalamus but from higher centers in the brain. We will have much more to say about the role of attention in subsequent chapters.

Some of the third-order thalamic neurons involved in the perception of pain have axons that project to a specific area of the cerebral cortex, and determining the significance of this connection was one of the great discoveries in modern neuroscience.

ATTRIBUTION: THE SENSORY CORTEX

Wilder Penfield and Theodore Rasmussen in the 1950s were aware of the general function of neurons in the cortex and were trying to identify sites that triggered epileptic seizures. To do so, they exposed the patient's brain by first removing a portion of the scalp and skull under anesthesia. When the anesthesia wore off and the patients awoke, they used tiny electrodes to stimulate the cortical circuits throughout each hemisphere (this is possible because the brain is insensate). There was little or no response when they probed most areas. Located at the approximate midpoint of each hemisphere is the *central sulcus*, and posterior to it is the *postcentral gyrus* (fig. 3.4). When they stimulated small areas of the cortex in the postcentral gyrus, the subjects reported feeling localized sensations. Remarkably, the responses from each locale along the gyrus were similar in all patients and were perceived as coming from the opposite side of the body. When they matched each response to the site of stimulation on the gyrus, they generated a distorted *somatotopic map* of the body

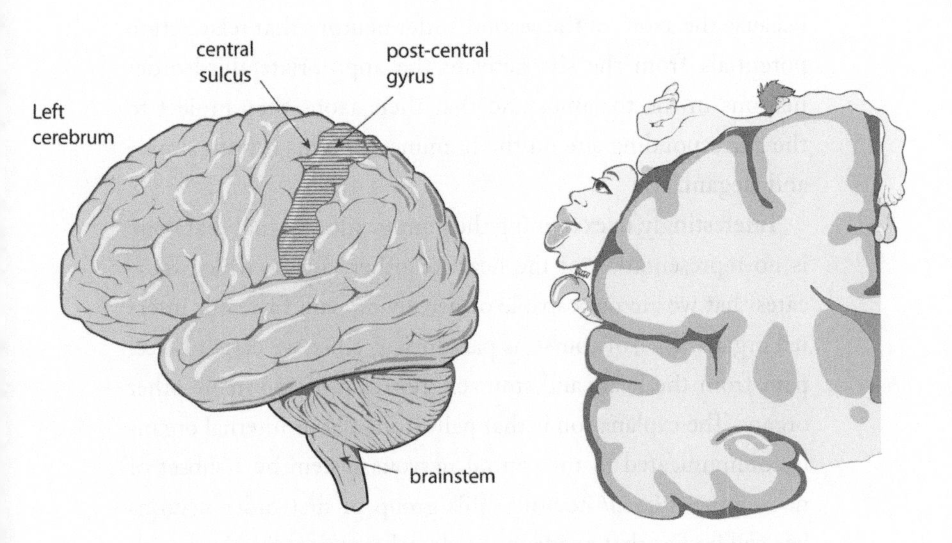

FIGURE 3.4 Left: The left cerebral hemisphere and the location of the central sulcus and the postcentral gyrus (hatched). Stimulation along the gyrus elicits reproducible sensations from the body. Right: A section through the postcentral gyrus in the left hemisphere showing the sensory homunculus that was created by correlating the origin of the sensation with the site of the stimulation along the surface of the postcentral gyrus. A similar homunculus is present on the right side.

that could be represented as a *sensory homunculus* (fig. 3.4).[3] There was evidence from studies of lower animals that such maps of the body existed, but to find it in humans was a major advance in our understanding of sensory processing.

At first glance, the distortion of the body map is perplexing, but subsequent studies showed that the map was an accurate reflection of the density of the inputs from each body region. The face, hands, and feet have an outsized representation in the cortex because these areas have the greatest number of inputs from first-order nerve endings. The representation of the fingers is especially large reflecting the sensitivity necessary for using the hand to manipulate objects. Most important is that we now know that the brain recognizes the source of a lesion because the axons of the second-order neurons that relay action potentials from the site activate the appropriate third-order neurons in the thalamus and that their axons then project to the corresponding site on the homunculus. This is both simple and elegant.[4]

Interestingly, in examining the homunculus we find that there is no representation of the heart or other viscera, which indicates that we are not capable of perceiving pain from our internal organs. This, of course, is paradoxical, since we certainly feel pain from the heart and stomach, not to mention many other organs. The explanation is that pain from all our internal organs is communicated to the central nervous system by a subset of nociceptive *visceral* neurons. This group of first-order neurons has cell bodies that reside in the dorsal root ganglia, along with those that mediate sensations from the dermatomes, but their peripheral processes reach their target organs via a unique set of nerves and the signal communicated by their central branches is processed differently. We will discuss these important neurons and their role in pain in chapter 7.

THE SOMATOSENSORY SYSTEM

If we connect the basic nociceptive pathway shown in figure 3.1 to the third-order neurons in the thalamus and their communication to the cortical neurons responsible for the homunculus, we create a *somatosensory system* that is responsible for us perceiving the pain from an injury and for localizing its source. Figure 3.5 shows the anatomical relationships of these components and illustrates that axons of the second-order nociceptive neurons from every spinal cord level cross over and ascend to the third-order neurons in the thalamus in a bundle that is known as the *spinothalamic tract*. The name is easy to remember because it defines a pathway that connects neurons in the spinal cord to neurons in the thalamus. Damage to the spinothalamic tract will have an effect on pain signals coming from the opposite side of the body, which is important in diagnosing spinal cord injuries.

THE PINPRICK MODEL

We can put all of this information in a more relevant context by considering a practical example—a simple pinprick to the right index finger. The pin pierces the skin, resulting in the generation of action potentials at the terminals of the first-order nociceptive neurons in the area of the injury.[5] Since we know that a branch of the median nerve supplies this finger, the potentials will propagate along the peripheral and then the central processes of the nociceptive neurons within the branches of this nerve. The central branches enter the dorsal region of the spinal cord at levels C5 to T1, where each bifurcates. One branch activates (indirectly) motor neurons whose axons course within the median nerve to elicit contraction of the muscles for the

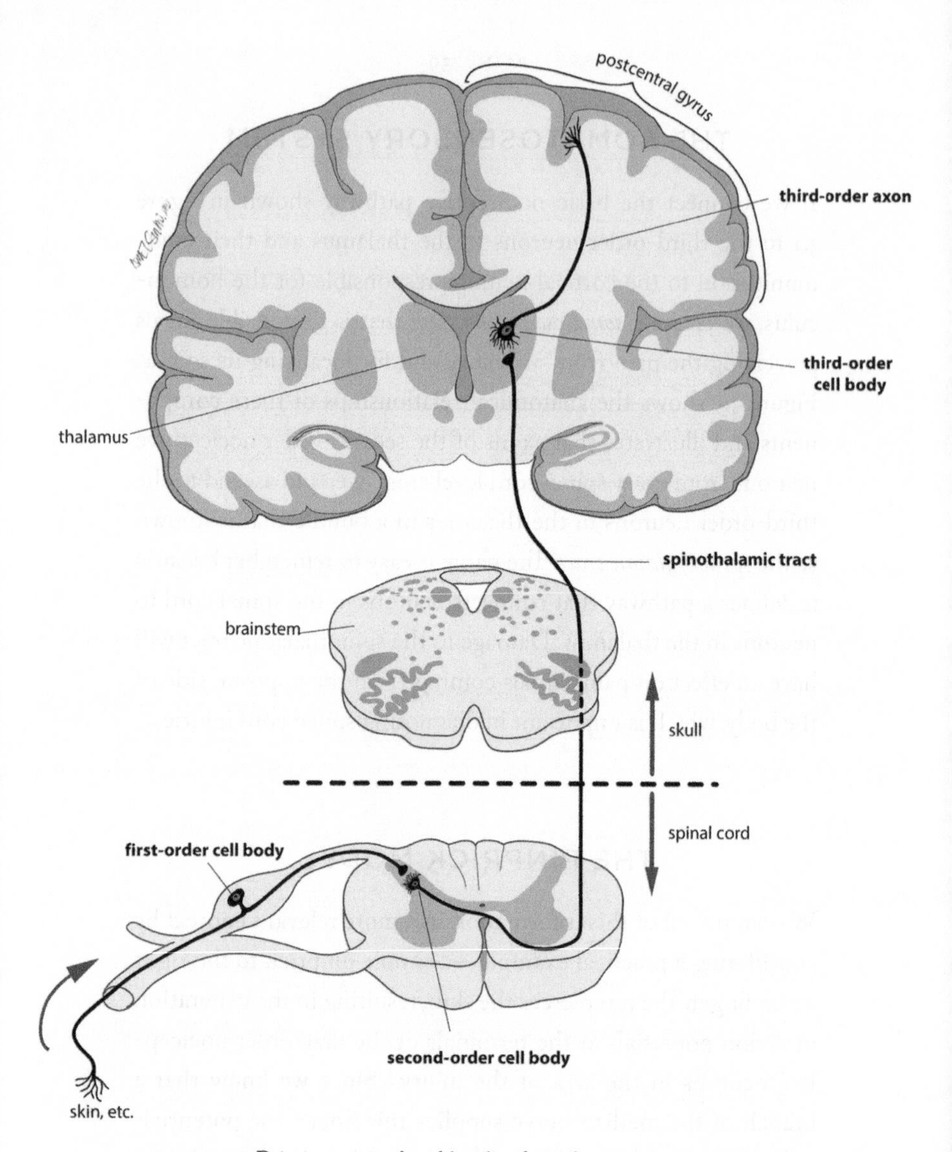

FIGURE 3.5 Pain is perceived and localized via the somatosensory system. The figure shows a first-order C-type nociceptive neuron with a peripheral process that innervates the skin, a cell body in the dorsal root ganglion, and a central process that activates a second-order neuron in the spinal cord. The axon of the second-order neuron crosses to the opposite side and ascends within the spinothalamic tract to the thalamus, where it activates a third-order neuron whose axon communicates with neurons in the sensory cortex.

defensive withdrawal of the finger, thereby protecting it from further injury. The other branch communicates via a synapse with the dendrites of the second-order nociceptive neurons. This elicits action potentials in the second-order neurons whose axons cross and ascend in the spinothalamic tract to the *left side* of the brain, where they activate the third-order neurons in the left thalamus where the pain is perceived. Some of the thalamic neurons will send axons to the region of the sensory homunculus in the left cerebral hemisphere that will attribute the pain to the right index finger.

Think about this for a minute: we are now aware that we have an injury to our right index finger merely due to the inputs to the brain via the pathways of the somatosensory system (see fig. 3.5). Moreover, parallel pathways will mediate the response to a pinprick from anywhere on the skin: only the dermatome and its nerve will be different. While this is correct, we have only described lesions on the body from the neck and below. What we need to discuss now is how pain information is communicated from regions in the head.

INNERVATION OF THE OROFACIAL REGION OF THE HEAD

The sensory homunculus contains a very large contribution from the facial region that reflects its importance for survival. The nociceptive pathway that sends information about a lesion from the face is communicated by cranial nerves that are functionally comparable to the spinal nerves we have discussed in detail. Like the spinal system, the cranial pathways also consist of first-, second-, and third-order neurons, but their anatomical disposition is different from their spinal counterparts below. The difference begins when the segmented spinal cord becomes the brainstem at the foramen magnum, which is the large opening at the base of skull. From this point upward there are no dorsal root ganglia, and the origin and distribution of the peripheral afferent and efferent nerves is much more complex. Fortunately, we don't have to discuss this anatomy in detail because the vast majority of nociceptive information from structures in the head is mediated by the first-order neurons whose cell bodies reside in the pair of very large trigeminal ganglia (fig. 3.6). These ganglia are located outside the brain stem and house sensory neurons that have functions comparable to those in the dorsal root ganglia. As the name indicates, each ganglion has three main branches that ramify extensively throughout the oral and facial region as the counterparts of a spinal nerve. Contained within each branch are the peripheral processes of the first-order nociceptive neurons that will convey signals for pain via its three branches. Each branch has unique targets, with no overlap, and collectively provides information from the teeth, gums, the nasal region, the tongue and oral region, the eye, ear, and the membranes surrounding much of the brain. Thus, the territory supplied by each trigeminal nerve is extraordinarily extensive and encompasses some of the most important structures in the body.

The central processes of the first-order nociceptive neurons enter the brainstem where they synapse on second-order neurons whose axons cross to the opposite side and ascend to synapse on the third-order neurons in the thalamus. The axons of the third-order neurons then project to the sensory homunculus in the territory that corresponds to the face. The cranial nociceptive pathway is identical to the spinal pathway except that the second-order axons communicate with a different subset of third-order thalamic neurons than those that receive inputs

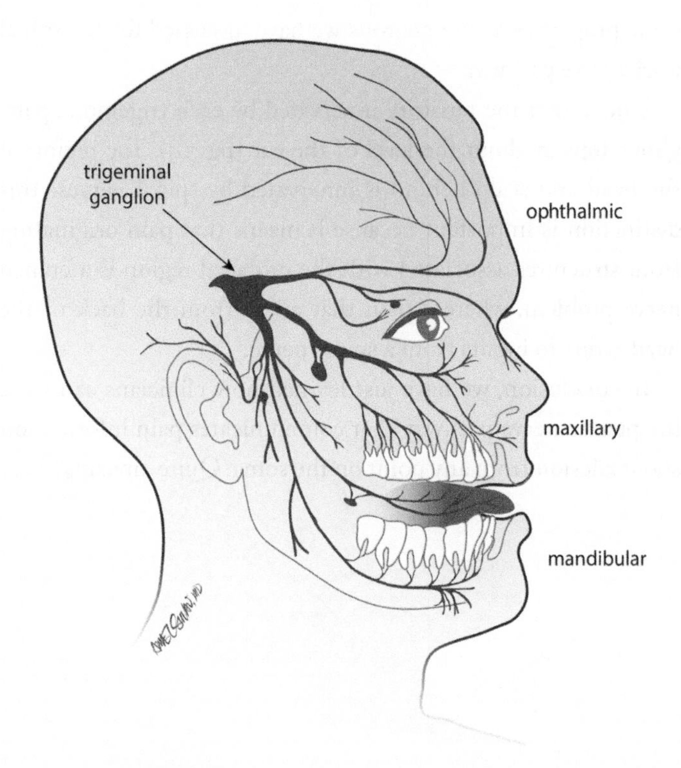

FIGURE 3.6 Pain from the orofacial region is initiated by first-order cell bodies that reside in the pair of large trigeminal ganglia. The figure shows the right ganglion and its ophthalmic, maxillary, and mandibular branches, each of which conveys sensory information from unique targets in its territory. Note that the branches only innervate structures, including the scalp, in front of the ear.

from the spinal nerves (fig. 3.3). We therefore have two different but parallel pathways for pain: one from the orofacial region and the other from the rest of the body. Most important is that pain from trigeminal neuralgia (or tic douloureux), migraine headaches, sinusitis, otitis media, toothache, etc. is communicated by first-order, second-order, and third-order neurons that have the same properties as the neurons we have discussed for the spinal nociceptive pathway.

Notice that the territory innervated by each trigeminal ganglion stops at about the level of the ear (fig. 3.5). The region of the head and scalp behind is innervated by spinal nerves. This distinction is important because it means that pain originating from structures associated with the orofacial region is a cranial nerve problem, whereas pain that arises from the back of the head is due to inputs from a spinal nerve.

In conclusion, we have just learned how clinicians can trace the precise nerve pathway that communicates pain information about a lesion from any point on the soma. Quite amazing!

4

THE MOLECULAR NEUROBIOLOGY OF PAIN

Let's briefly put what we have learned in perspective. The neuroanatomists of the late 1800s were aware that peripheral nerves were responsible for sensations and motor movements, but despite painstakingly determining the distribution of every nerve, they had little understanding of how the nerves actually functioned. A major advance occurred with the advent of the microscope and the discovery of neurons. The progression from macroscopic anatomy to microscopic cellular neurobiology eventually resulted in determining that nociceptive information was communicated via the sequential activation of the first-, second-, and third-order nociceptive neurons that comprise the nociceptive somatosensory system. Nevertheless, knowing the pathway of communication did not begin to explain how this system actually mediates the signals for pain. This understanding only came about in the latter part of the twentieth century with the ascendancy of molecular biology. Now, for the first time, neuroscientists could attribute specific neuronal functions to mechanisms at the molecular level. To truly understand pain, therefore, we must look at these mechanisms in detail; we will devote the next several chapters to what we call the molecular basis of pain. As we proceed, we will also explain how the discoveries in molecular neuroscience provided the pharmaceutical industry with targets for the development of drugs to diminish pain.

In general, the first-order neurons respond to three different events that result in the perception of pain. *Noxious pain* results from an event that pierces or damages the skin and the underlying tissue. This includes a cut or burn. *Neuropathic pain* results from an injury that severs, crushes, or otherwise damages a major nerve and the axons within. This is usually caused by a very serious injury and the repair process is long and complex. The final type of lesion that causes pain is an inflammation, and *inflammatory pain* can be present in the absence of an obvious noxious or neuropathic involvement. Although each type of lesion activates first-order C-type neurons, each has unique elements at the molecular level. We will discuss each of these lesions in turn, beginning by focusing on the events that underlie noxious pain.

RECEPTORS AND CHANNELS: ACUTE NOCICEPTION AT THE LESION SITE

If we exclude the special senses, such as vision and hearing, we learn about the external word via the endings of the peripheral axons of sensory neurons that are located in the skin of a dermatome and underlying structures. The endings (or terminals) for touch, vibration, and other modalities are enclosed within specialized structures that transduce the stimulus into action potentials (fig. 4.1). The peripheral terminals of the first-order C-type nociceptive neurons are different in that they are "naked," meaning they are directly exposed to the interstitial space of surrounding tissues.[1] Each terminal is an enlarged region of the peripheral process that is bounded by a membrane that forms a

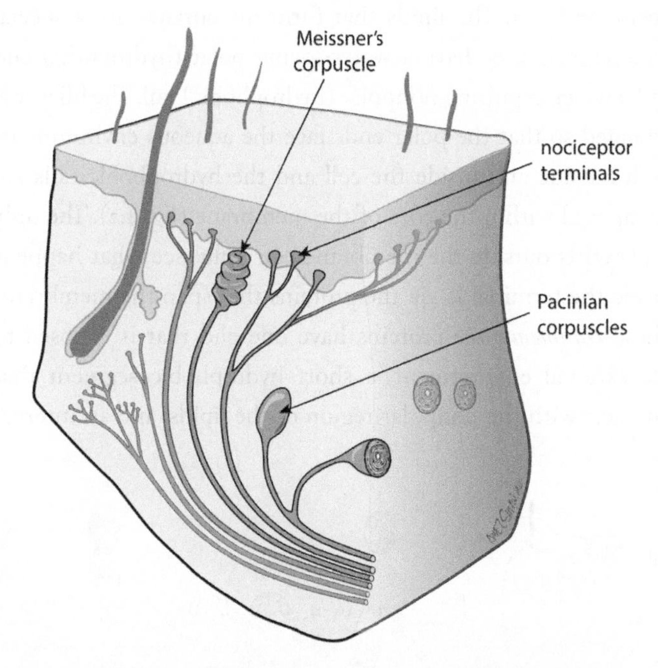

FIGURE 4.1 Three-dimensional view of the skin showing the terminations of the peripheral processes of sensory neurons. The terminals for touch are encapsulated in Meissner's corpuscle and those that detect vibrations are within Pacinian corpuscles. In contrast, nociceptive nerve endings are "naked" and directly exposed to the space in surrounding tissues.

barrier separating the external aqueous environment from that inside the terminal. How then does an external event, such as an injury, activate the nociceptive neurons? The process whereby information about an external event is communicated to the interior of a cell is known as *signal transduction*, and understanding how this occurs is a prerequisite for understanding pain.

We know that biological membranes are comprised of a lipid bilayer. Lipids are commonly known as fats and we use them every day in soaps and detergents to facilitate the removal of grease and dirt. The lipids that form membranes are a special type because they have a water-loving polar (hydrophilic) end and a water-repulsing nonpolar (hydrophobic) tail. The bilayer is arranged so that the polar ends face the aqueous environments both outside and inside the cell and the hydrophobic tails are juxtaposed within the core of the membrane (fig. 4.2). The only way events outside the membrane can influence what happens inside the terminal is via the proteins that span the membrane. These *tranmembrane* proteins have one end that is exposed to the external environment, a short hydrophobic segment that interacts with the nonpolar region of the lipids, and an internal

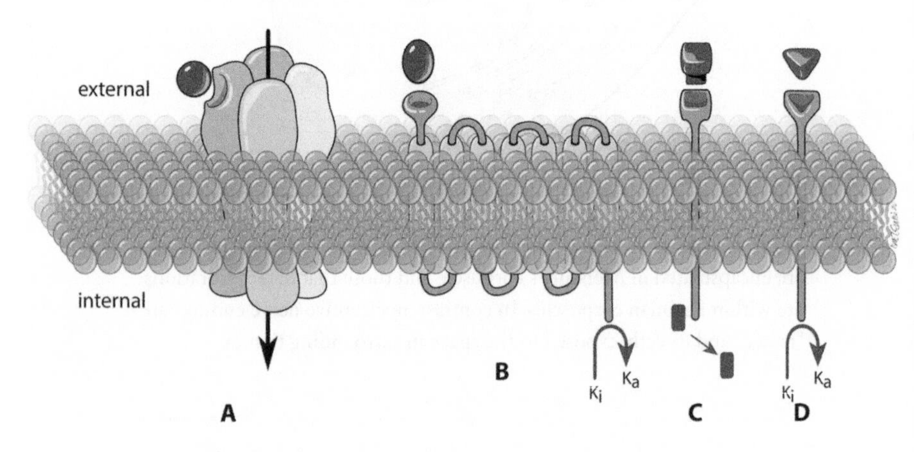

FIGURE 4.2 Trans-membrane receptor proteins span the lipid bilayer membrane to allow communication between the extracellular space and the inside of the neuron. Signal transduction occurs when the binding of a ligand (circle, etc.) to a site on the external surface of the receptor elicits a response inside the cell. (A) A receptor-gated ion channel in which the ligand opens the central channel (arrow) to allow ions to flow across the membrane. (B) A complex receptor that spans the membrane multiple times. (C, D) Simple receptor types. Binding of the ligand can activate a kinase (Ki to Ka) or the release of an enzyme (rectangle). The activated kinases and released enzymes can then migrate to mediate events elsewhere in the cell.

end within the terminal. Some are simple and traverse the membrane only once, whereas others are highly complex and can span the membrane as many as seven times. The external segment is folded into a three-dimensional structure containing a pocket that recognizes a small molecule known as a *ligand*. The analogy is that the ligand is a key and the pocket is the keyhole. When a ligand binds to the pocket it causes an immediate conformational change in the internal segment of the protein that will have important consequences. The organization of the membrane lipids with their polar groups exposed to aqueous environments outside and inside and the hydrophobic region in between gives membranes a fluid property. Hence, transmembrane proteins are essentially floating in a lipid sea, and unless their internal segment is anchored to components within the cell, they can migrate in the plane of the membrane and join with other proteins to form complexes.

Two classes of transmembrane proteins are present at the nociceptor terminal. The first, *ion channels*, are complexes of several subunits embedded in the membrane with a central channel or pore. The pore spans the membrane and thereby interconnects the aqueous environment outside with that inside. Each ion channel selectively regulates the passage of calcium (Ca^{++}), sodium (Na^{+}), or potassium (K^{+}) ions through the pore.[2] The pore is normally closed, but it opens in some channels in response to the binding of a ligand (see fig. 4.2a). The binding results in a rapid flow of ions through the channel (an influx) that continues until the ligand is released and the channel closes. The ligands and channels involved in regulating the entry of Ca^{++} are especially important because they initiate many events that are essential for pain. The channels for sodium and potassium are different because they open or close in response to changes in voltage: these voltage-gated channels are far more complicated and will be discussed later in this chapter.

The other class of transmembrane proteins at the terminal is the receptors responsible for the process known as signal transduction. Signal transduction is initiated when the ligand binds to its specific site on the external sequence of the receptor. This induces a conformational change in the internal segment of the receptor that triggers numerous events within the terminal (fig. 4.2C, D). Most common is the activation of an enzyme known as a *kinase*. Over five hundred kinases are known, and each transfers the terminal phosphate group in adenosine triphosphate (ATP) to a site on the kinase's target protein. This seemingly simple reaction, known as phosphorylation, is significant because the addition of the phosphate alters the function of the target.[3] Thus, signal transduction, via a ligand and receptor, allows each cell to marshal an appropriate internal response to a specific external change in its environment.

The terminals of the nociceptive neurons have many receptors, each of which responds to the presence of a specific extracellular ligand. Each receptor is also linked to a specific intracellular kinase that will be activated when the ligand binds to the receptor. Phosphorylation of a channel by a kinase, for example, can either increase or decrease the number of ions that can enter or exit, depending on the circumstances. Protein kinase A (PKA), protein kinase C (PKC), and protein kinase G are a few of the many kinases with an important role in generating the signals for pain. Most are not suitable targets for the development of analgesics, however, because they are present in many cell types throughout the body and blocking their function would have many serious side effects. We will only discuss kinases that are specifically associated with pain and whose distribution among other cell types is limited.

THE PINPRICK MODEL REVISITED

To understand how receptors and channels in the terminal membrane interact to initiate the signals for pain, let's revisit the simple pinprick model we discussed in chapter 3 (fig. 4.3). Keep in mind that the pinprick is to the right finger and that it just pierced the skin to cause some damage to the underlying tissue. We already know that this minor injury will elicit an immediate defensive withdrawal of the finger to avoid additional damage, followed by an acute (sharp) pain that we recognize as coming from the site of the puncture. The pain rapidly diminishes and is then forgotten. What has happened is that the tip of the needle has ruptured some of the cells just beneath the surface resulting in the release of their contents into the extracellular space. Among these is adenosine triphosphate (ATP). ATP is formed by the energy-driven addition of a phosphate group to adenosine diphosphate. This occurs in mitochondria for which the fuel is glucose, or in chloroplasts, for which the energy comes from the sun. The energy "stored" in ATP can then be used to drive reactions elsewhere in the cell. As mentioned earlier, kinases are the catalysts for these reactions. Since thousands of reactions occur in cells each second, ATP is one of the most abundant constituents, so its presence in the interstitial space is a natural marker for cell injury. Thus, the binding of ATP to its receptor on the terminal membrane is a proximal source of the signals that will ultimately be felt as painful.

To be more precise, the binding of ATP to its receptor initiates signal transduction events inside the terminal that ultimately evoke action potentials (electrical impulses) that are the primary means of communication between neurons and their targets. As described in chapter 3, the action potentials will propagate rapidly along the components of the somatosensory system. Since

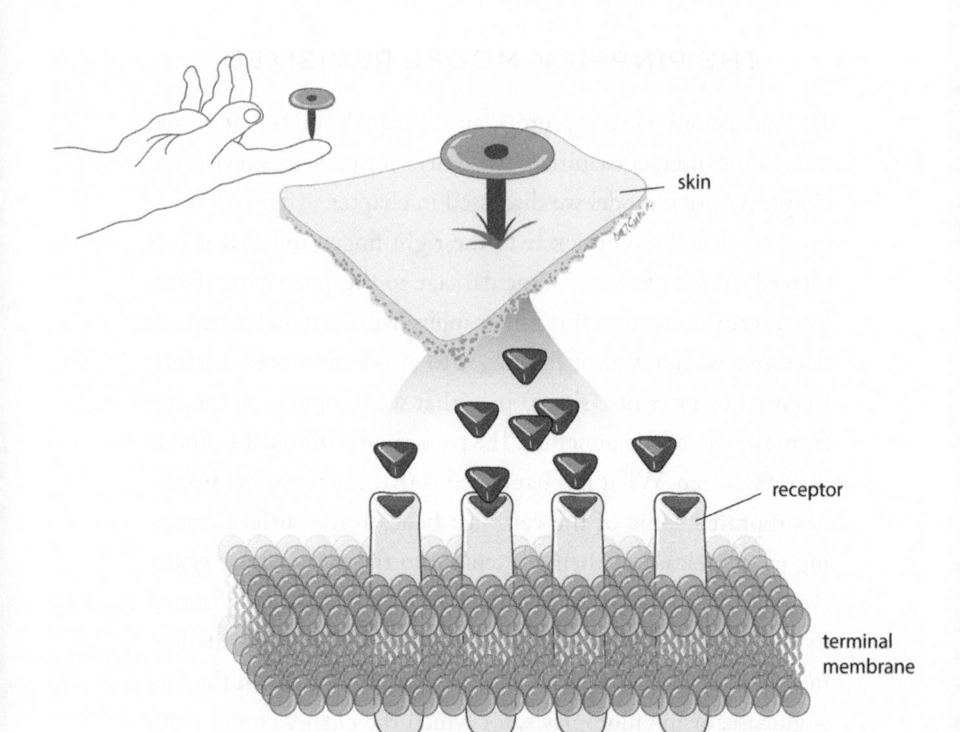

FIGURE 4.3 The simple pinprick model. A pin pierces the skin and ruptures underlying cells, resulting in the release of ATP (triangles) that binds to its receptor on the membrane of the peripheral terminal. The binding alters the conformation of the internal segment of the receptor thereby activating kinases and other enzymes within the terminal (curved arrows). These signal transduction events will alter the properties of the terminal and will result in the opening ion channels and the generation of action potentials.

the axons of the second-order neurons cross to the opposite side of the spinal cord, they will activate third-order neurons in the *left* thalamus, which will then communicate with the *left* cerebral cortex. Thus, the pain from the pinprick to the right finger will be perceived in the *left* thalamus and will be attributed to the right index finger via circuits in the *left* homunculus.

ACTION POTENTIALS AND THE INTENSITY OF PAIN

Still considering the pinprick model, it will become obvious that a very important piece of information is missing. Just how does the circuitry in the brain perceive the pain from the pinprick to be relatively mild as opposed to the pain from a more serious injury that will be much more intense? The answer is that *the intensity of pain is encoded in the number of action potentials that are elicited at the injury site*. Given their importance we need to look at the generation of action potentials in more detail.

We can describe the events responsible due to the efforts of Alan Lloyd Hodgkin and Andrew Fielding Huxley, who presented, in 1952, a model that explains the ionic mechanisms underlying the initiation and propagation of action potentials. Their pioneering studies utilized the squid giant axon; many other advances in neuroscience came from studies of so-called simple invertebrate systems. They received the 1963 Nobel Prize in Physiology or Medicine for this work. Action potentials are generated by the movement of ions through the channels in the terminal membrane (fig. 4.4). There is a great difference in the concentration of K^+ and Na^+ ions between the inside and outside of the terminal. The concentration of K^+ inside is roughly twenty times that found in the external space whereas the concentration of Na^+ is roughly fifteen times higher outside than inside. This creates an electrical potential difference of −70 millivolts (mV) inside versus the outside that is known as the resting or equilibrium state. There is also the concentration difference, and the net result is that Na^+ is poised to pour through its channel to enter

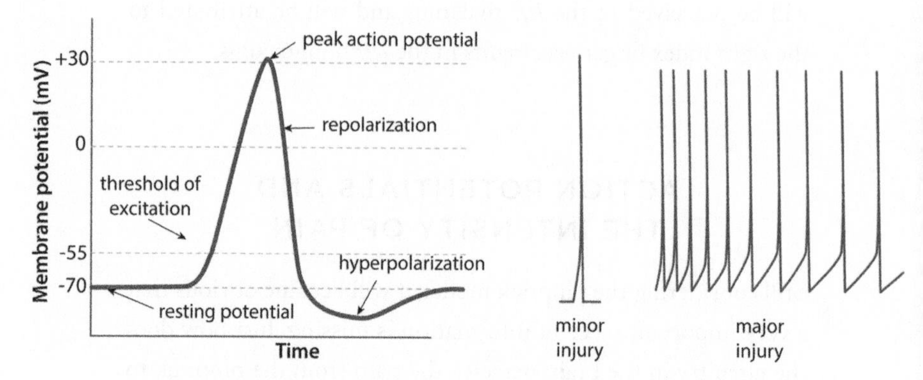

FIGURE 4.4 Left: An action potential. The resting membrane potential within the terminal is -70 mV. After an injury, a signal transduction event results in an influx of Ca^{++} that leads to the opening of some Na^+ channels. When the change in potential reaches a threshold, many more Na^+ channels open, depolarizing the terminal and resulting in the rising phase of the action potential. At the peak amplitude of +30mV, the Na^+ channels abruptly close and K^+ channels open. The efflux of K^+ results in the falling phase of the action potential that leads to a hyperpolarization and then a restoration to the normal resting potential. Right: A minor injury elicits a single action potential whereas a serious injury elicits many action potentials.

the terminal whereas K^+ is poised to exit through its channel. They are only poised because under resting conditions the gate to each channel is essentially closed and we feel no pain. When ATP or other noxious agent from an injury binds to its receptor, calcium channels open allowing Ca^{++} from the external space to enter the terminal. The entry of Ca^{++} opens the gate in some of the sodium channels and the resulting influx of Na^+ begins to make the inside of the terminal more positive. If the influx of Na^+ passes a certain *threshold* level, such that the potential goes from -70 to -55 mV, then more Na^+ gates open flooding the terminal and resulting in a rapid depolarization. Electrophysiologists can record these changes in potential by implanting tiny electrodes into the tissue and monitoring the change with an oscilloscope. The Na^+ influx appears as a rapid rise in the membrane potential (see fig. 4.4). Once the influx exceeds the threshold, the process cannot be halted and the result will be a full action potential. This is referred to as the *all or none law*. The influx of Na^+ eventually causes the interior potential to reach approximately +30 mV, which is the maximum height of the action potential and is known as the *amplitude*. As this occurs, the Na^+ gates begin to close and K^+ gates open, driving potassium to the outside and making the interior potential negative again. This creates the falling phase of the action potential. The fall actually reaches a state in which the inside potential reaches minus 80 mV, which is a hyperpolarization. The duration of the hyperpolarization is important because another action potential cannot be elicited during this period. The K^+ gates close and with the assistance of a Na^+/K^+ pump, equilibrium is restored. All of this occurs in a few thousandths of a second.[4]

Na^+ and K^+ channels are not restricted to the terminal but are also distributed along the peripheral and central processes of the first-order nociceptive neurons. Because these channels respond to changes in the resting potentials, (i.e., they are voltage gated), the generation of an action potential at the terminal will open sodium channels in the adjacent neuronal process. This will result in the generation of an action potential in that region, which will then activate the adjacent region, and so on.[5] For each action potential generated at the peripheral terminal, a corresponding action potential, at the same amplitude, will be generated in the processes, and they will propagate all the way to the spinal cord. Traveling at a rapid rate, it takes only a few hundredths of a second for the signal from the terminal in the skin to reach the central nervous system. Such speed is essential to communicate the occurrence of an injury.[6] Like electricity flowing through a wire, action potentials provide a way to rapidly communicate information from the periphery to the spinal cord and brain.

Most important for us to remember is that the information about the injury is not encoded in the amplitude, which does not decrease as it travels along the processes of the first-order nociceptive neurons, but rather in the number of action potentials. Naturally, the intensity will also depend on how many first-order neurons are activated.

SYNAPSES AND THE INJURY RESPONSE

We know from our discussions that the central processes of the first-order neurons bifurcate within the dorsal horn of the spinal cord. One branch then communicates with the neurons that mediate the rapid reflexive withdrawal of the finger. The other activates the second-order nociceptive neurons whose axon ascends to the thalamus resulting in a localized pain of short duration. Of course, we now understand that the action potentials traveling along the central processes are responsible for both responses. However, action potentials are merely electrical signals, and to understand how they affect the target we must look at the structure and function of *synapses*, which are present at the point where the central processes communicate with their targets. Each central process ends as a presynaptic terminal that is located close to the postsynaptic terminal on the dendrites of the target (fig. 4.5). The space between the two endings is known as the synaptic cleft.

The combination of the presynaptic and postsynaptic terminals and the cleft forms a *chemical synapse*.[7] Studies beginning

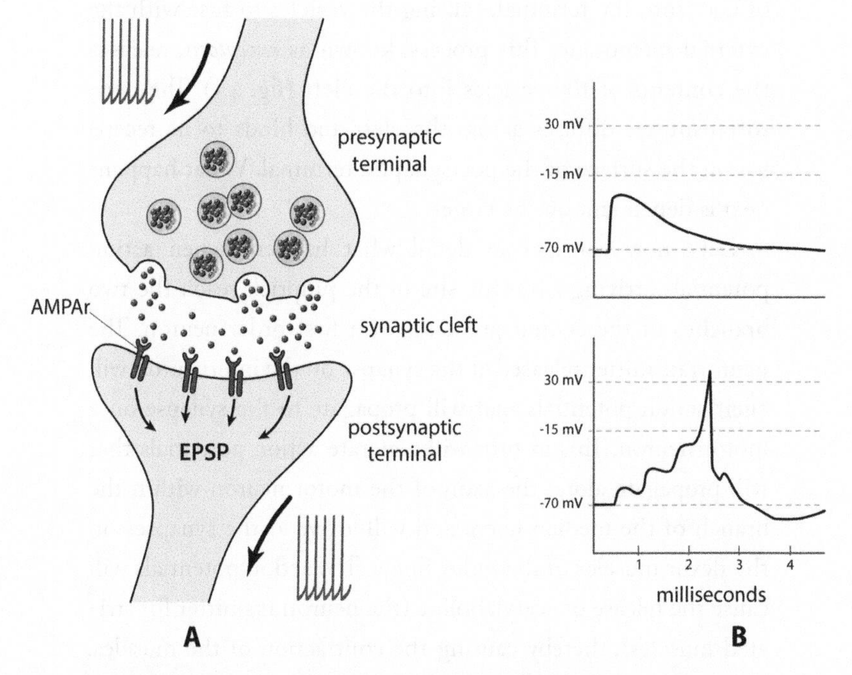

FIGURE 4.5 The synapse: structure and function. (A) Diagram showing the presynaptic terminal with vesicles containing neurotransmitter, the synaptic cleft, and the postsynaptic terminal with receptors in its membrane. (B) Top: An excitatory postsynaptic potential (EPSP) in the postsynaptic terminal Bottom: Three EPSPs have summated to generate an action potential.

in the last half of the twentieth century have provided a highly detailed description of the morphology of the synapse and the molecular events that mediate synaptic function. The presynaptic terminal is much larger in volume than the axon and contains a population of small membranous vesicles that are filled with a neurotransmitter. There are many different neurotransmitters that define the specific function of the synaptic connection. In all, the arrival of action potentials triggers the release of Ca^{++} into the terminal, causing the vesicles to fuse with the external membrane. This process, known as *exocytosis*, releases the contents of the vesicles into the cleft (fig. 4.5). The neurotransmitter diffuses across the cleft and binds to its receptor on the surface of the postsynaptic terminal. What happens next is dependent on the target.

Let's now consider in detail what happens when action potentials arriving from the site of the pinprick reach the two branches of the central processes of a first-order neuron. The neurotransmitter released at the synapse on the interneuron will elicit action potentials that will propagate to the synapse on a motor neuron. This in turn will generate action potentials that will propagate along the axon of the motor neuron within the branch of the median nerve and will arrive at the synapses on the flexor muscles of the index finger. The action potentials will cause the release of acetylcholine (the neurotransmitter for striated muscles), thereby causing the contraction of the muscles. This is the pathway by which the pinprick maintains the protective withdrawal of the finger.

The central process also synapses on a second-order nociceptive neuron, and one might reasonably ask why this synapse is even necessary. After all, the branches of the central process of the first-order neurons could just continue directly to the brain. The answer is that the transmission of information across a chemical synapse can be regulated and this is especially true here. In fact, the synapse between the first- and second-order neurons is an essential site for the control of pain because it is the gateway to the thalamus. To begin to understand how this occurs, let's look at the function of this synapse after the pinprick.

The vesicles in the presynaptic terminals of the first-order neurons contain *glutamate*, which is the primary neurotransmitter for nociception. When the action potentials from the site of the pinprick arrive they mobilize the vesicles, which then fuse with the external membrane and release the glutamate into the cleft. The glutamate diffuses to the membrane of the postsynaptic terminal of the second-order neuron, where it is recognized by *AMPA receptors* (fig. 4.5). These receptors are different from the traditional receptors we have discussed because they are activated by a ligand, glutamate, not voltage, and have an ion channel built into their structure. Such receptors are termed ionotropic. *AMPA receptors are the primary mediators of nociception.* Binding glutamate to the receptor directly opens the channel and results in an influx of Na^+. Unlike the voltage-gated channels that can remain open for relatively long periods, the channel in the AMPA receptor opens and closes very quickly depending on the concentration of glutamate. The flow of Na^+ into the terminal increases the membrane potential (i.e., increases the positive charge) and generates what is known as an *excitatory postsynaptic potential* (EPSP) (fig. 4.5). The size of the EPSP across any given AMPA channel is dependent upon the concentration of glutamate, but EPSPs within the entire terminal are additive and when they reach a threshold, an action potential is generated in the second-order neuron. The action potential will then propagate to the third-order neurons in the thalamus resulting in the perception of pain from the pinprick and, via circuits in the sensory homunculus, the attribution of the pain to the region of the index finger that was pricked. Note that the number of incoming action potentials from the injury site will correlate with the amount of glutamate released and the number of outgoing action potentials to the brain. Thus, the few action potentials from the site of the pinprick will be translated into a minor pain of short duration. What happens in response to a serious injury is fascinating but more complicated and will be discussed in more detail in the next chapter.

SODIUM CHANNELS

Although action potentials can be generated via the binding of glutamate to its ionotropic AMPA receptor, the most important channels in nociception are the voltage-gated sodium channels that are responsible for the propagation of action potentials along the axon. The realization that sodium channels are essential for pain came about only in the latter part of the twentieth century, but they were actually the unknown targets of agents that had been developed many decades earlier and which have become a part of the culture of pain management.[7] We have all heard of Novocaine, which was synthesized by Alfred Einhorn in 1905, and proved to be a remarkably effective local anesthetic. Injection of Novocaine resulted in a loss of sensation around the site that lasted for approximately thirty minutes without ill effects on consciousness or cognition. Many other "caine" derivatives soon followed, but the initial hope that these agents would be used to prevent pain during surgery was never realized because the anesthesia was short-lived and surgeons preferred to use general anesthetics. Nevertheless, agents such as Lidocaine were a boon to dentistry because they made drilling cavities pain-free.

All of the variants of Novocaine are lipophilic, and they exert their effect by blocking sodium channels. The correlation between the empirical findings that agents such as Lidocaine are effective anesthetics and the evidence that they act exclusively on voltage-gated sodium channels was exciting because it indicated that developing a long-lasting sodium channel blocker would be a panacea for victims of persistent pain. As happens with many searches for a "magic bullet," complications have made this search far more difficult than anticipated.

The core of the problem is that there are many different types of voltage-gated sodium channels.[8] A prototypical sodium channel is comprised of alpha and beta subunits (fig. 4.6). The alpha subunit is extraordinarily complex because the protein weaves its way back and forth across the membrane many times to create multiple domains. Each alpha unit contains an ion-conducting channel located adjacent to a voltage-sensing region and multiple sites on the internal loops that can modulate channel activity. The beta subunit regulates the overall function of the channel. Not only is the structure complex, but the human genome has the code for nine α subunits.

What this means is that there are at least nine different subtypes of voltage-gated sodium channel, each with unique properties and responses. Neuroscientists have designated these NaV1.1–NaV1.9. Remember that different agents released or produced in response to injury will activate different kinase pathways that selectively phosphorylate unique sites on the channel. In theory,

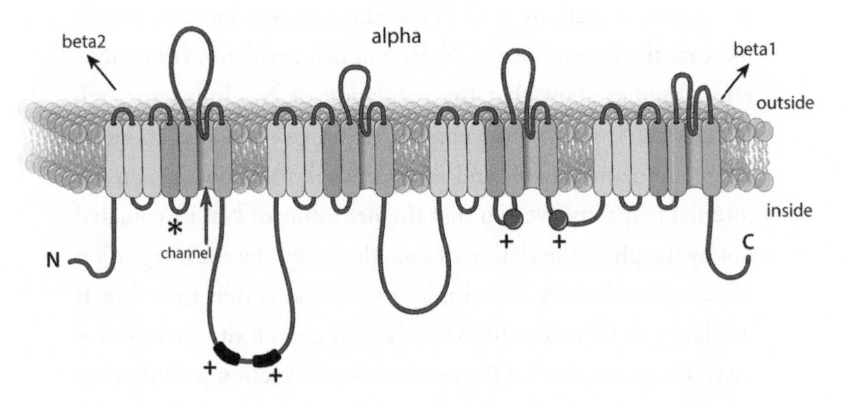

FIGURE 4.6 The alpha subunit of a voltage-gated sodium channel. The protein crosses the membrane many times creating loops both outside and inside. It interacts with beta subunits and contains the channel through which the Na^+ enters the cell adjacent to a voltage-sensing region (*) and several sites on the inner loops (+) that can modulate channel activity. Note that both ends of the protein are located on the inside, which differs from the transmembrane receptor proteins (see note 8 for additional details).

these sites could be targeted to block a particular channel and whatever pain it signals. NaV1.7 and NaV 1.8 are the focus of such efforts because both are found in the C-type nociceptive neurons and NaV1.7 is localized at their terminals. NaV1.8 is a unique channel because it is resistant to tetrodotoxin (TTX)[9] and its phosphorylation by kinases in response to various agents has been linked to an increased perception of pain.

Whether developing a site-specific inhibitor of phosphorylation in either subtype is a worthwhile approach is an open question because there is conflicting evidence as to which subtype is most important for pain. The strongest case can be framed for NaV1.7 because there is substantial clinical evidence that it is involved in human paroxysmal pain disorders. Moreover, congenital mutations that result in a nonfunctional NaV1.7 completely eliminate the sensation of pain so that stimuli such as contact with sharp objects or burning hot materials fail to elicit any painful sensations at all. While the studies of human patients indicate the importance of NaV1.7 in pain, evidence from laboratory studies show that the regulation of Na^+ influx through the channel is much more complicated than previously thought. There are many kinases and many phosphorylation sites on the internal loops involved so that the net influx of Na^+ is regulated not by the phosphorylation of a single site but by the integration of phosphorylations at multiple sites: blocking one, therefore, is not likely to be successful. Why is such complexity then necessary? The answer is that the generation of an action potential has many consequences and must be tightly controlled. Requiring multiple steps to activate the channel means that no single event is enough.

5

ADAPTATION

ADAPTATION MODULATES PAIN

The previous chapters described the relatively simple molecular mechanisms involved in the response to a pinprick to the index finger. All of the responses occurred seconds after the injury, and the pain diminished relatively rapidly because the injury was minor. The consequences of a more serious injury are naturally more dire because the pain is more intense and of longer duration. One might guess that more severe, longer lasting pain would involve a different group of nociceptive neurons, but this would not be correct. The same first-, second-, and third-order nociceptive neurons that contribute to acute pain are also responsible for persistent pain; the circuits in the thalamus and cortex are also the same. How is it possible that such a relatively simple system can ensure that the intensity and duration of the pain are commensurate to the severity of the injury? The answer is that a process known as adaptation modulates the response to pain. Adaptation is a form of *neuronal plasticity*,[1] which is the inherent ability of the nervous system to adjust to events.

The adaptive modulation of pain perception occurs at both the peripheral terminals and at the synapse between the first- and second-order neurons. It is regulated at both sites by processes that are intrinsic to the neurons in the pathway. There are also very important external influences as well that will be discussed in chapter 8. To understand what happens at the terminals, let's revisit the simple analogy of a switch that controls the current in a wire that enables us to turn on and off a light bulb. If we want to get more control of the light, we can modify the switch into a rheostat so the light can not only be turned on or off but also dimmed or brightened. By analogy, the severity and duration of pain can be modulated simply by altering the membrane threshold of the terminal so as to increase or decrease the number of action potentials elicited. In essence, adaptation changes the perception of pain by regulating the function of the receptors and channels that are intrinsic to the first- and second-order neurons.

RESPONSES AT PERIPHERAL TERMINALS TO A MORE SERIOUS INJURY

A great deal of effort has been devoted to understanding the molecular mechanisms that underlie adaptation because in many ways chronic pain can be viewed as adaptation gone awry. Consider a serious cut to the right index finger that has caused considerable damage to the underlying tissues. The level of ATP released will be much higher than that after the pinprick because more cells have been destroyed. We can therefore make several predictions. First, more action potentials will be generated. Second, they will propagate to the spinal cord where they will both evoke the finger withdrawal reflex and elicit many action potentials in the second-order neurons. Third, these signals will reach the thalamus and result in a perception of pain that is more intense than that after the pinprick. Lastly, the signals forwarded to the homunculus will attribute the injury is to the right index finger. All this should be familiar because these pathways are the same as those described after the pinprick. However, there is one notable difference: the duration of the pain will increase markedly because the finger has sustained significant damage and an awareness of the injury must be maintained to protect the finger until the healing is complete. Thus, the pain will persist for hours or perhaps a day or longer, not just minutes or hours as was the case with the pinprick.. How the neurons in the pain pathway adapt to the seriousness of an injury is interesting and complex.

In addition to more ATP being released into the surrounding area (see fig. 5.1), the cut will trigger the release of compounds from the excited terminals themselves. Two of these, *calcitonin gene–related peptide* (CGRP) and *substance-P* increase the vascular permeability of nearby blood vessels resulting in the swelling and redness that occurs in the area around the injury. Mast cells recruited from the circulation will synthesize precursors of the prostaglandins that will have important roles in the initiation of pain signals. These are direct consequences of the cut. There will also be an invasion of the site by cells of the immune system whose role is to neutralize potential pathogens and remove debris from damaged tissue. Nerve growth factor (NGF) and other compounds released from these cells will result in an inflammation, so that a veritable cocktail of molecular agents will be present in the space around the terminals of the first-order neurons (fig. 5.1). We will discuss the most important inflammatory mediators of pain in chapter 7. Some of the agents might even diffuse to activate the terminals of nearby nociceptive neurons that were not directly impacted by the injury. This will amplify the response because more neurons will send signals to the spinal cord and will also expand the area that is painful. We will now focus on the two agents that are significant potential targets for the development of analgesics.

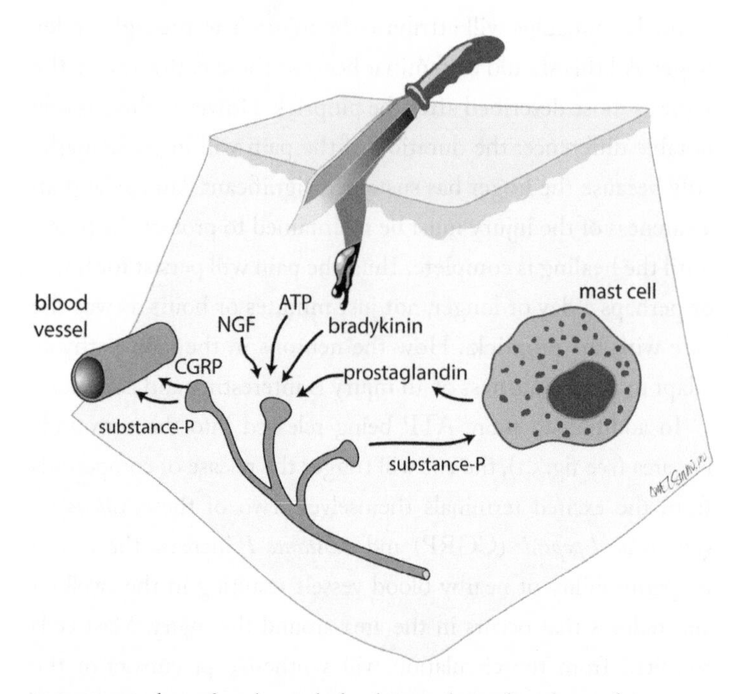

FIGURE 5.1 A cut deep beneath the skin results in the release of agents from the terminals of the nociceptive neurons and from immune cells recruited to the site. These agents will dramatically alter the locale around the injury and elicit action potentials that will propagate to the spinal cord and brain.

Bradykinin

Bradykinin has multiple roles in initiating the signals for pain. It is a nonapeptide (i.e., a chain of nine amino acids) that is released by the cleavage of a large plasma protein at the injury site.[2] The binding of bradykinin to its transmembrane receptor on the surface of the terminal results, as expected, in the initiation of an enzyme cascade within the terminal. The cascade results in the activation of kinases that then catalyze the transfer of the terminal phosphate of ATP to sites on the voltage-gated Na^+ channels. This simple reaction has profound consequences because it lowers the threshold of the channels so that they now generate action potentials in response to a stimulus that would not normally have an effect. Put another way, the terminal has been *sensitized*.[3] The phosphorylation of the Na^+ channels is an example of a *posttranslational reaction*. Translation is the mechanism that assembles amino acids into a protein; this occurs in the neuronal cell body. Posttranslational therefore refers to the modification of an already existing protein, which can occur anywhere in the neuron. However, posttranslational changes are transient because they can be readily reversed. For example, phosphatases can remove the phosphate from the channels and receptors, thereby restoring the system to the noninjured state.

In addition to its direct role in sensitization, the binding of bradykinin to its receptor activates enzymes known as phospholipases. These enzymes cleave phospholipids in the membrane to produce arachadonic acid, which is then released into the space around the terminal. It is taken up by mast cells and modified by an enzyme known as *cyclooxygenase* (COX) to produce *prostaglandins* (fig. 5.1). Prostaglandins are notorious for exacerbating pain by contributing to the sensitization. The COX enzyme is inhibited by derivatives of salicylic acid. One such derivative, acetylsalicylic acid, is commonly known as aspirin, probably the most effective and widely used analgesic in the world. Its analgesic properties were actually discovered a thousand years ago when ancient physicians used extracts of willow bark to alleviate pain. Willow bark contains salicylic acid and the willow tree belongs to the genus Salix. We can now understand why we take aspirin to relieve pain.

Nerve Growth Factor

The other key agent involved in adaptation is nerve growth factor (NGF), another small peptide that is also cleaved from a larger precursor and is present in many different cell types. Its name is somewhat of a misnomer because it was first recognized for its ability to promote neuronal growth during embryonic development, but this role is superseded in adults, where it contributes to pain signaling.[4] Indeed, cutaneous administration of NGF leads to pain within one to three hours and the pharmaceutical industry has targeted NGF for the development of analgesics. When released into the area of the injury, NGF has several profound short- and long-term effects on the generation of signals for pain. All these effects are mediated by the membrane receptor (TrkA), which is selectively expressed on the peripheral terminals of the first-order nociceptive neurons and on the mast cells that were recruited to the site of the injury. Binding of NGF to the TrkA receptor on mast cells causes the release of histamine and serotonin (5HT) that also contribute to the pain from an inflammation.

Taken collectively, the events mediated by bradykinin and NGF result in what is known as *peripheral sensitization*, an important form of adaption that partially explains two important pain-related phenomena. The first is *allodynia*, which is defined as a painful response to a stimulus that would normally not cause pain. For example, a mere touch to skin that would not elicit pain will be very painful if applied in the area of the pinprick or cut. The second is *hyperalgesia*; a stimulus that would normally cause some pain results in enhanced pain, e.g., pressing on an injured area. Thus, allodynia and hyperalgesia reflect the ability to fine-tune the level and extent of pain depending upon the severity of the injury.

How long peripheral sensitization lasts is an open question. The most likely answer is that it is transient. Studies have shown that the effects of bradykinin last only minutes, and many of the important changes that are initiated by the injury at the terminal are the result of posttranslational changes. One example we mentioned was the posttranslational phosphorylation of the sodium channel by kinases. Posttranslational changes are short-lived and are readily reversible due to the presence of phosphatases that remove the phosphates and restore the channel to its original (normal) resting level of activity. The ability to initiate an event via a kinase-mediated phosphorylation and terminate the event via a phosphatase-mediated dephosphorylation is one way in which cells can rapidly adjust to changing conditions.

NOT ALL PAIN IS THE SAME

Thus far we have explained how the signals from a lesion are perceived as pain, but pain can differ in quality and in the temporal order in which the signals reach the brain. We need to refine our understanding of pain by considering once again the serious injury to the index finger, but this time we will focus on the nature of the pain. The first response will be an acute pricking, stabbing, or lancing pain that is highly localized. This is followed by an intense pain that is more widespread and perceived as a dull burning, throbbing, or aching. As we have discussed, the first type of pain diminishes rapidly, responds to readily available analgesics, and is usually not of great concern. The second type is far more serious and has been attributed to the C-type neurons. This is not to say that we have a complete understanding of how events in the periphery give rise to the various types of pain. Just recently, for example, a network of glial cells was detected just beneath the skin. Most glial cells are intimately associated with neurons, but these seem to be independent and have an important role in the development of pain from mechanical injuries, such as from a crushing blow. More work is necessary to understand how this occurs, but since these glial cells do not have any connections to the CNS, they must function by exerting an influence on the terminals of the first-order nociceptive neurons. We will again focus on the C-type neurons because they are activated by the events that result in peripheral sensitization and are most likely to be important for understanding persistent and chronic pain. In addition, individuals who lack this nociceptor type have a congenital absence or reduced sensitivity to pain. Finally, the C-type neurons respond to chemical and thermal stimulation and transmit information for the burning type of pain. Burning pain is often associated with chronic conditions, and we will therefore look the mechanisms that are involved.

THERMAL PAIN

The survival of an animal depends on its ability to detect and avoid situations that can cause tissue damage. In the previous chapters, we discussed responses to injuries that have pierced the skin and damaged underlying tissues. Extremes of temperature can also damage tissues and cause burning pain. How we perceive temperature is an intriguing problem that fascinated Greek and Roman philosophers and it is only relatively recently that we have identified the key components involved. It turns out that we sense gradations in temperature via a family of transient receptor protein vanilloid (TRPV) channels that are present in the terminal membrane of sensory neurons.[5] When these channels open in response to changes in temperature, an influx of ions into the terminal results in the generation of action potentials. The number of action potentials increases as the temperature increases. Some family members respond specifically to temperatures in what we consider to be the normal range, which extends to about 43°C (109°F). Two others are most important here because they are activated only by temperatures that can cause tissue damage. The TRPV1 responds in the 40–50°C (104–122°F) range whereas TRPV2 responds to temperatures between 50–60°C (122–140°F). Forty-three degrees Celsius (109°F) is close to the pain threshold in humans and also to the threshold for the activation of C-type nociceptive neurons. The TRPV1 is present at the terminals of these neurons and has been studied extensively due to its contribution to burning pain in pathological pain states, such as fibromyalgia, and postherpetic neuralgia (shingles), and for its role in inflammation, which will be discussed at length in chapter 7.

The TRPV1, like the sodium channels that we discussed previously, is a complex of transmembrane proteins. Unlike the sodium channel, however, the TRPV1 is a heat-gated ion channel that opens at its temperature threshold to allow Ca^{++} to enter the terminal. We already know that this triggers the activation of sodium channels and the generation of action potentials. Moreover, as the temperature of the skin increases, the threshold for the activation of the TRPV1 channel decreases, meaning that the channel is becoming sensitized. Consequently, a temperature that would normally elicit a few action potentials resulting in only mild pain will now cause more intense pain because more action potentials will be generated. This *thermal hyperalgesia* is easy to understand if you remember that even mild heat applied to skin that has suffered from a sunburn will cause far more pain than would occur in the absence of the sunburn. Most of us have experienced the burning sensation elicited by eating chili peppers.[6] This is due to the presence of capsaicin, which is a naturally occurring ingredient in hot peppers and a direct (agonist) activator of TRPV1 channels. Capsaicin elicits an intense burning sensation in humans, but at high concentrations it rapidly *desensitizes* the TRPV1, i.e., it closes the channel gate. The exact mechanism is still being investigated, but clinicians have taken advantage of this unusual property to use capsaicin to relieve pain from a variety of conditions, including osteoarthritis, fibromyalgia, and peripheral neuropathies. Pain from minor injuries can often be relieved by applying a topical cream containing low amounts of capsaicin.

In addition to activation of the TRPV1 in response to a direct burn, thermal hyperalgesia also occurs after an injury; we are well aware that exposing an injury site to even mild heat causes pain. Humans have evolved in a way that provides alternative mechanisms for responding to serious threats. Having backup systems is a good idea because they increase the chances of survival. It is not completely surprising, therefore, that the TRPV1 is also activated by bradykinin and NGF. Studies have shown that injecting bradykinin into human skin produces a dose-dependent pain and heat hyperalgesia. The mechanism is exactly what we would predict: the bradykinin binds to its receptor on the terminal membrane, thereby activating a kinase that phosphorylates the TRPV1, altering its structure such that the gate that controls the entry of Ca^{++} now opens at a lower temperature and more action potentials will be generated.

The function of NGF is more complicated because it influences the function of the TRPV1 in two ways. The first involves the binding of NGF to its TrkA receptor, followed by the now familiar activation of a kinase, the phosphorylation of the TRPV1, and the increased entry of Ca^{++}. The second is more novel and is based on the idea that located inside the terminal are small vesicles that have TRPV1 in their membrane. In response to an injury, the binding of NGF to the TrkA results in the movement of the vesicles and their fusion with the external membrane of the terminal that both increases the number of TRPV1 channels on the surface and results in more action potential discharges. Which of the two mechanisms is more important is not clear. What is clear is that thermal hyperalgesia is an important concomitant of persistent burning pain and that the TRPV1 is a central component in this process. Consequently, developing a drug to block the TRPV1 is the focus of many efforts to prevent or at least alleviate this type of pain. A major impediment is that the TRPV1 is found in many tissues, such as the urinary tract, bladder, gastrointestinal tract, and in many areas of the central nervous system. Its function in these tissues is not known, but any drug that inhibits the TRPV1 is very likely to have significant side effects. Another problem is that the drug would have to target the TRPV1 exclusively or else it would interfere with the function of the other TRPV family members and disrupt responses to non-noxious temperatures. As might be imagined, designing drugs with high target specificity is very difficult.

ADAPTATION IN THE SPINAL CORD

The two most important parameters of pain are its intensity and duration. We have already discussed at length how intensity is encoded by the number of lesion-elicited action potentials. How the nervous system regulates the duration of the pain is more complex and has obvious implications for chronic pain. We can be certain that peripheral sensitization plays a role in maintaining an awareness of pain because it allows action potentials to be elicited by minimal stimuli long after the initial barrage that was generated immediately after the injury. This explains how merely touching (allodynia) or compressing (hyperalgesia) the cut finger will result in pain. However, peripheral sensitization provides only a partial explanation for how the pain from the cut finger lasts longer than the pain from a pinprick.

What studies have shown is that the expression of hyperalgesia and allodynia also require alterations in the electrophysiological properties at the synapse between the first- and second-order neurons. Like the changes in the periphery that sensitize the peripheral terminal, these alterations sensitize the synapse so that the number of action potentials that arrive from the injury site is increased in the second-order neurons, resulting in more activation of the thalamus. Two terms are used when referring to this adaptive potentiation. *Central sensitization* is the broader term because it includes all changes in nociceptive neurons without distinguishing between regulatory processes intrinsic to the neurons and those that are imposed via external sources. The ability of extrinsic circuits to regulate the pain pathway is very important and will be discussed in subsequent chapters. The other term, *long-term potentiation* (LTP), refers to an increase in synaptic strength that, in the context of pain, is due to changes that occur at the synapse between first-order C-type neurons and their second-order neuronal targets. LTP is especially important for the expression of allodynia and is the term that will be used here.

LONG-TERM POTENTIATION

What we will describe in the next few paragraphs took neuroscientists decades to decipher and required the development of highly sophisticated instruments and molecular probes, somewhat like the development of the microscope and stains that exposed the world of neurons. What was found is that LTP can be divided into an early and late phase and that each has a specific role in the duration of pain. The early phase rapidly sensitizes the synapse and acts in conjunction with peripheral sensitization to explain the initial allodynia and hyperalgesia. The late phase prolongs the sensitization by actually altering the protein composition of the terminal. To understand how these changes occur, let's examine the function of the synapse in greater detail.

We know from our discussion of the pinprick model that the initial barrage of action potentials will cause the release of glutamate from the presynaptic terminal. The glutamate will bind to the ionotropic AMPA receptors on the postsynaptic membrane and, if the subsequent depolarization is sufficient, will generate action potentials in the second-order neuron. The result will be immediate pain from the pinprick. But how do we explain that we can still feel pain minutes or even hours afterward, especially when the area is touched? The answer is that the extended pain occurs via the activation of NMDA receptors in the early phase of LTP.

NMDA receptors are also present in the membrane of the postsynaptic terminal (fig. 5.2). Like the AMPA receptors, they are ionotropic, but they differ because the channel in the NMDA has a preference for Ca^{++} rather than Na^{+}. In addition, the channel is blocked by a tightly bound magnesium ion (Mg^{++}). The Mg^{++} block can be removed by an intensive depolarization of the postsynaptic terminal, which is what happens in response to the initial barrage of action potentials from the pinprick.[7] The removal of the Mg^{++} blockade allows an influx of Ca^{++} that initiates the activation of kinases that phosphorylate membrane-bound receptors and ion channels. The summation of all these events results in the terminal being more receptive to incoming action potentials, i.e., it is *sensitized.*

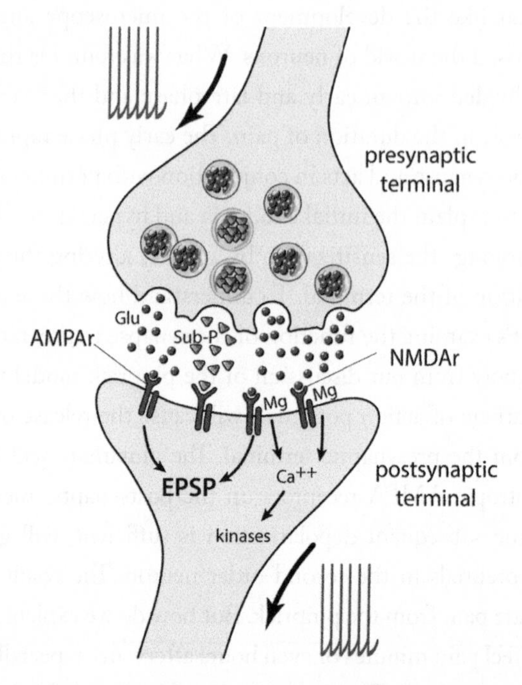

FIGURE 5.2 Long-term potentiation at the synapse in response to a serious injury. The barrage of action potentials arriving at the presynaptic terminal from the injury site results in the release of glutamate and substance-P and the activation of their respective receptors on the membrane of the postsynaptic terminal. The resulting depolarization is sufficient to relieve the Mg^{++} block, thereby activating the NMDA receptors. The subsequent influx of Ca^{++} through the receptor pore initiates changes that sensitize the terminal.

Given what we have just learned, let's consider what happens when the area of the pinprick is gently touched. Normally this would not elicit any action potentials and cause no pain, but because the peripheral terminal is sensitized, a few action potential potentials will be elicited. When they reach the synapse, they will be amplified because the postsynaptic terminal has also been sensitized: more action potentials will be generated and the gentle touch will now be painful, thereby explaining how allodynia keeps us aware of the injury. However, remember that the allodynia is maintained by posttranslational modifications that are transient, which means that it will only last minutes or hours,

Of course, the deep cut to the finger is a much more serious injury that will generate many more action potentials than the pinprick. The result will be much more intense pain that will extend for many hours or even days. The prolonged duration is attributed largely to the events that occur in the late phase of LTP. Surprisingly, this prolongation of pain involves many of the same players as the early phase of LTP except that the much more intense barrage of action potentials from the cut will cause a persistent activation of the NMDA receptors. This, in turn, will activate several kinases within the postsynaptic terminal that will profoundly change the electrical properties of the second-order neurons. Some of the kinases enter the nucleus where they turn on genes in the DNA that ultimately results in the synthesis of new ion channels and receptors that are subsequently inserted into the postsynaptic membrane. Thus, the late phase of LTP occurs because these newly synthesized proteins have altered the properties of the terminal to enhance the magnitude of the response. Because these events involve the activation of DNA in the genome, they are known as *phenotypic changes*, which, unlike the transient posttranslational changes, cannot be easily reversed.[8] Thus, in theory, the sensitivity and pain due to the late phase of LTP can potentially last indefinitely.

We have just learned that the activation of NMDA receptors is responsible for the transition between the transient pain of a pinprick and the longer duration of the pain from more serious injuries. One other point should be made, however, and that concerns the finding that some of the first-order nociceptive neurons release peptide neurotransmitters in addition to glutamate. Substance-P, for example, is released in the periphery after an injury (fig. 5.1), and it is also released from the presynaptic terminal to activate its NK1 receptor on the postsynaptic membrane (fig. 5.2). There is evidence that the binding contributes to the generation of EPSPs and thereby contributes to the activation of the NMDA receptors. There are also indications that substance-P can interact with other cells in the CNS, where its exact role is not clear. We only mention this because drugs are being developed to block the effects of substance-P, but the results have been inconclusive.

Since we know quite a lot about the molecular and electrophysiological events responsible for the late phase of LTP, it would be wonderful if this information could be used to understand the mechanisms underlying persistent or chronic pain. Unfortunately, experiments in model systems indicate that the duration of LTP does not correlate with the duration of pain that persists for a very long time. Thus, the two phases of LTP ensure that the duration of pain is commensurate with the seriousness of the initiating lesion, but they cannot explain how pain can last for weeks or longer. What we have learned, however, is that the most effective way to extend the duration of the pain is to change the phenotype of the nociceptive neurons. In fact, recent studies have identified two other alterations that can prolong pain, perhaps by extending the late phase of LTP, and these are the next topics to be discussed.

6

MOLECULAR SIGNALS FOR PERSISTENT PAIN

RETROGRADELY TRANSPORTED SIGNALS REGULATE GENE EXPRESSION

We have a high degree of confidence that peripheral sensitization and the two phases of long-term potentiation (LTP) can explain how pain from most injuries is maintained from the initial awareness of the injury throughout the hours or days required for healing. Pain that lasts a week or more, however, is a different and far more serious matter. We just learned that one way to extend the duration of pain is to alter the neuronal phenotype by activating the genome. Thus, the late phase of LTP prolongs pain because the function of the second order nociceptive neurons has been changed due to the addition of newly synthesized kinases, receptors, and channels. However, we have recently learned that other phenotypic changes can occur in nociceptive neurons and that their appearance correlates with long-lasting pain. These changes are slower in onset, involve molecular signals that are activated only by severe injuries, and can potentially transform the neurons indefinitely, making them especially relevant as sources of chronic pain. One involves a kinase that is only activated by serious injuries, and the other results in the synthesis of proteins that impact synaptic transmission between the first- and second-order neurons.

To understand the mechanisms involved, we first have to recognize that whereas the terminals are the working ends of the neuron, only the cell body contains the genome and machinery for the synthesis of proteins and other macromolecules. What this means in terms of nociception is that any changes in the composition of the terminals of the first-order neurons in the distant periphery are dependent on the cell body. This poses two significant logistic problems because the total volume of the peripheral process is thousands of times the volume of the cell body and the terminals are located vast distances from the cell body in the dorsal root ganglion.

There are two ways that macromolecules made in the cell body reach the terminals. Soluble proteins and other constituents move within the peripheral processes via a mechanism known as *slow axoplasmic flow*. This occurs at a rate of about 5 mm/day, which means that it can take weeks before a constituent made in the cell body will reach a terminal in the periphery. The other mechanism is *fast axonal transport*, which is responsible for moving all membranous components, including channels, receptors, and vesicles containing neurotransmitters and other contents at a rate of 400 mm/day. Both the flow and transport occur constantly to replenish constituents at the terminal that have reached their allotted lifetime. But the lifetime of the proteins at the terminals correlates with the activity of the terminal itself; the greater the activity, the greater the need for replacements. The findings by neuroscientists that synapses can be shaped by experience, as during learning and memory, suggested that there must be molecular messengers that travel from the periphery to the cell body where they direct the genome and manufacturing centers to synthesize constituents needed at the terminals. This process is highly relevant to nociception, where nothing much happens at the terminals under normal conditions, but a great deal occurs following an injury. There is a mechanism by which certain proteins in the axon and at the terminal can be rapidly transported back (retrogradely) to the cell body. We call them *sentinel proteins* because they monitor the integrity of the axon and terminals and can direct the genome to respond to any changes. Rapid transport, in this case, is estimated to be 200 mm/day, which means it will take hours or even days for these retrogradely transported signals to reach the cell body and activate the genome; additional time is then needed for the newly synthesized components to be transported to the terminal. Consequently, these signals are activated in first-order neurons only following very severe injuries, where the pain awareness must last for many days, weeks, or longer: they are the ultimate mechanism responsible for persistent pain because they change the properties of the nociceptive neurons after traumatic injury. The retrograde transport system itself is a marvel comprised of molecular motors that transport cargo along tracks within the axon. We do not fully understand all the molecular processes involved in this transport, nor do we know how many retrograde signals there are. So far two such signals have been identified and their mode of action has provided important insights as to how persistent pain is regulated.

THE INDUCTION OF A LONG-TERM HYPEREXCITABILITY

Studies of conditions such as cystitis, osteoarthritis, colitis, and metastatic bone cancer have shown that persistent pain is associated with the presence of a *long-term hyperexcitability* (LTH) in the cell bodies of the first-order neurons. This hyperexcitable state is manifest as a lowering of the threshold for the generation of action potentials. Like the LTP at the synapse, this means that even a single action potential from the lesion site that reaches the cell body of these neurons will generate multiple action potentials, which will then propagate along the central process to the synapse on the second-order neurons and then to the thalamus and homunculus. LTH is an amplification system due to changes in the electrophysiological properties of the cell body, not the terminal or synapse. Most important, it is a phenotypic change that we know can last indefinitely. Since the LTH will sustain both allodynia and hyperalgesia as long as it is present, even a light touch or mild heat applied to the area of the lesion will be painful (fig. 6.1). This type of hypersensitivity is characteristic of many types of chronic pain.

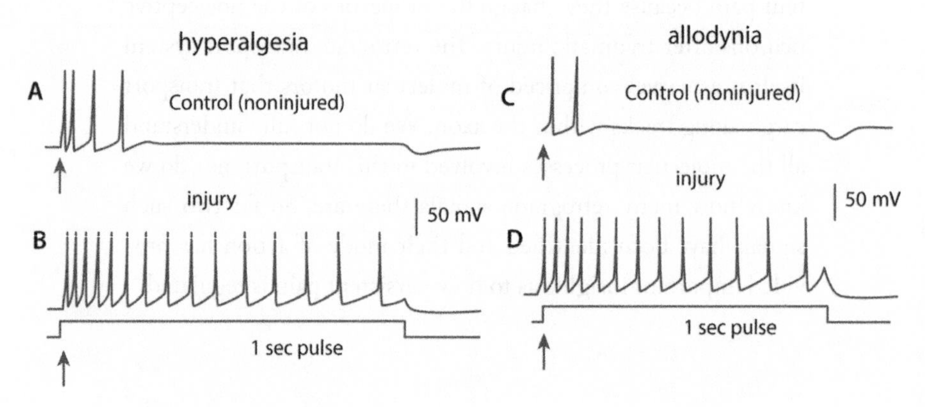

FIGURE 6.1 Long-term hyperexcitability enhances action potential discharge contributing to hyperalgesia and allodynia. The excitability of first-order neurons was evaluated by electrically stimulating the cell body (arrow) to elicit action potentials. (A) Several action potentials are elicited in a non-injured neuron responding to a mildly painful event. (B) An identical stimulus to a neuron expressing LTH due to an injury elicits many more action potentials. (C) Two action potentials are elicited in a neuron in response to touching the skin. (D) A comparable touch to an area of injured skin elicits many more action potentials in a neuron exhibiting LTH.

LTH has obvious implications for the treatment of chronic pain. One significant characteristic that has emerged from studies is that it appears only after a considerable delay and so cannot have any role in acute pain or the pain mediated by the early phase of LTP. The delay occurs because the sentinel protein activated at the site of the injury has to use the retrograde transport system to get back to the cell body of the affected neurons.

A plausible and intriguing possibility is that the LTH extends the duration of the late-phase LTP. To explore this possibility required understanding the events responsible for the appearance of the LTH, which meant identifying the sentinel protein. This was a major challenge because, as we know, even the simplest vertebrate nerves contain hundreds of axons and thousands of proteins. My group at Columbia University circumvented this problem by using the relatively simple nervous system of the marine mollusk *Aplysia californica* (fig. 6.2A). The utility of this organism's nervous system was pioneered by Dr. Eric Kandel and his colleagues at Columbia University in order to identify molecules involved in learning and memory. Naturally there was initial skepticism about using an invertebrate to investigate these very human attributes, but his choice was vindicated when he shared the Nobel Prize in Physiology and Medicine. The many advantages of using the *Aplysia* nervous system for the study of nociception include its very large neurons that can be recognized reproducibly from animal to animal so that the same neuron can be examined using different experimental protocols. Also important is that the nervous system can be removed from the animal and studied *in vitro* (out of the body).

We were particularly interested in *Aplysia* as a model to study the molecular neurobiology of pain due to the bilateral group of sensory neurons that responded to injuries to the body wall and appeared to be an invertebrate version of the C-type first-order

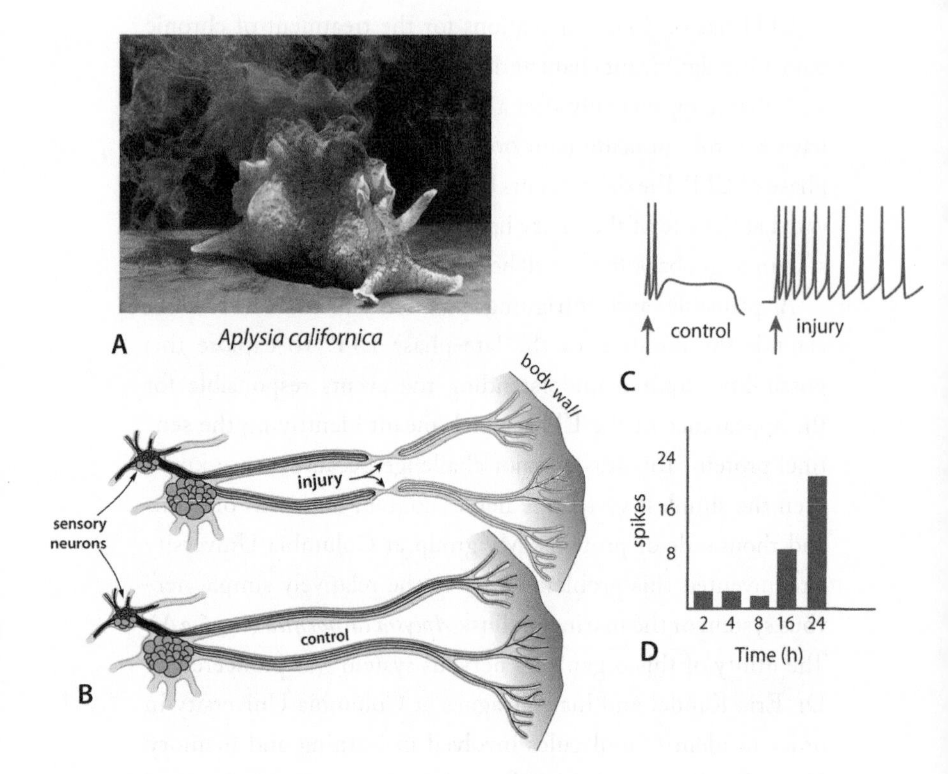

FIGURE 6.2 Development of LTH in *Aplysia* sensory neurons. (A) Photograph of *Aplysia californica* releasing ink in response to a noxious stimulus. (B) Part of the nervous system in isolation showing the location of the bilateral cluster of sensory neurons. To induce LTH, the nerves to the body wall that contain axons of the sensory neurons were crushed on one side (arrow). The nerves on the other side were untouched. (C) A representative result twenty-four hours later showing that stimulation of a cell body on the noninjured (control) side resulted in a few action potentials whereas an identical stimulus to a neuron on the injured side resulted in a barrage of action potentials. (D) A time course study shows that the increase in excitability appeared only after a twenty-four-hour delay, confirming the expression of LTH.

nociceptive neurons (fig. 6.2B).[1] We found that a serious injury to the axons of these neurons resulted in an LTH that appeared in the injured cell bodies but only after a delay that correlated with the time required for a sentinel injury signal to be transported back to the cell body (fig. 6.2C, D).

Subsequent studies showed that the appearance of the LTH required the synthesis of new proteins in the cell body. Taken together, these findings offered compelling evidence that we were seeing the *Aplysia* version of mammalian LTH. We were now much better positioned to test the hypothesis that the LTH is induced by an injury-activated sentinel protein. In addition to the relatively large size of the nociceptive neurons, axoplasm can be extruded from the axons for analysis and Dr. Ying-Ju Sung in my lab exploited these advantages to identify the protein responsible for the induction of the LTH as *protein kinase G-1α* (PKG).[2]

PKG: A MOLECULAR SWITCH FOR PAIN

This kinase is perfectly situated as a signal for persistent pain because studies of mammalian animal models showed that it is enriched in the axons of the C-type nociceptive neurons that exhibit both the late-phase LTP and LTH. Significantly, it is also absent from motor neurons, meaning that a drug that inhibits PKG will not affect muscle movement. PKG is one of the few sentinel proteins to be identified, and it was a challenge to determine how it works.

The process begins in the cell body where PKG is synthesized (fig. 6.3). It then enters the peripheral axon of the first-order neurons in an inactive form and migrates by slow axoplasmic flow that moves the mass of soluble proteins toward the terminals. When there is a severe injury or lesion, there is a brief but extensive

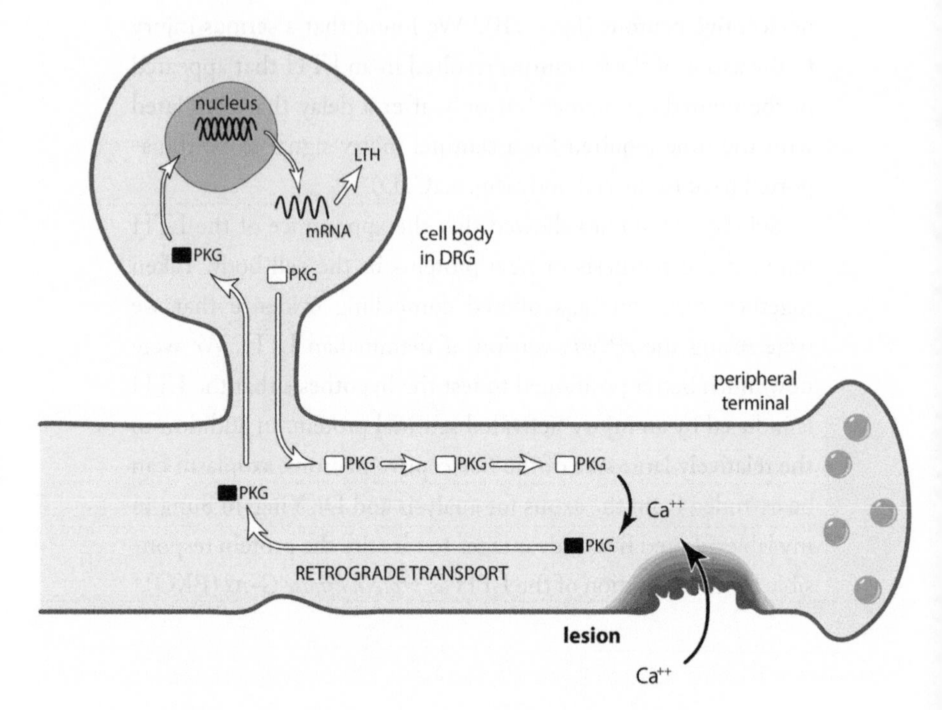

FIGURE 6.3 Retrograde transport of protein kinase G after a lesion to the peripheral axon. PKG is synthesized in the cell body and contains a concealed signal sequence (white box). The PKG enters the axon and migrates slowly toward the terminal. A lesion along the axon, or at the terminal, allows Ca^{++} to enter, which initiates an enzyme cascade that activates the kinase and exposes the signal sequence (now a black box). The exposed sequence is recognized by the retrograde transport system and the active PKG is rapidly conveyed to the cell body where it enters the nucleus. The PKG ultimately activates the genome resulting in the synthesis of proteins that are responsible for the LTH.

influx of Ca^{++} from the external space through the wound into the axon or into the peripheral terminals. The Ca^{++} initiates a specific enzyme cascade that causes a change in the three-dimensional conformation of PKG—the protein unfolds, with two consequences (fig. 6.3). First, PKG becomes enzymatically active;

second, the unfolding exposes a short sequence of amino acids that is normally hidden within the kinase. This exposed signal sequence is recognized by a component of the retrograde transport machinery and serves as a ticket for the rapid transport of PKG back to the cell body. In *Aplysia* the signal sequence provides direct entry into the nucleus, but in mammalian neurons the PKG phosphorylates another kinase, which then enters the nucleus. A very important outcome from these studies was that the LTH was induced in response to an inflammation as well as an injury, which indicates that PKG is a signal specific to nociception.[3] In addition, the fact that the LTH is a property of the cell body is significant because it means that the newly synthesized proteins do not have to be exported into the axon.

LTH is a concomitant of conditions of prolonged pain, and PKG is an essential factor in the induction of LTH. Consequently, we can view PKG as a molecular switch for persistent pain and a very attractive target for the development of analgesics for intractable pain. What makes it even more attractive is that PKG is only activated by severe injuries and has no role in acute pain. Consequently, an inhibitor of PKG will not prevent the pain that follows the usual types of minor injuries that are instructive for avoiding damaging events. The next challenge is to identify the proteins that are the targets of PKG. They are directly responsible for the LTH and would be even better targets for drugs to alleviate pain.

NERVE GROWTH FACTOR REVISITED

Another very important signal for persistent pain is nerve growth factor (NGF). We have already discussed the role of NGF in the development of peripheral sensitization and its importance in thermal hyperalgesia, but its long-term effects have drawn the most attention. Clinical studies have implicated NGF in many chronic pain conditions and there is evidence that NGF-mediated signaling is an ongoing and active process in chronic nociceptive pain states. We know that a variety of lesions result in the release of NGF into the spaces around the peripheral terminals. The long-term role of NGF occurs when the NGF binds to its TrkA receptor on the terminal membrane. Unlike its role in short-lived pain, this binding results in the internalization of a vesicle with the NGF-TrkA receptor facing the inside. The vesicle enters the retrograde transport system and is conveyed within the peripheral process to the cell body of the first-order nociceptive neurons in the DRG where it activates the genome to mobilize the neuron for pain signaling.[4] Among the changes is an increase in the expression of the cell surface receptor for bradykinin, as well as an increase in the number of voltage-gated sodium channels, voltage-gated calcium channels, and the TRPV1. NGF-TrkA signaling also leads to an increased expression of the neurotransmitters substance-P and calcitonin gene–related peptide (CGRP). Since we have already discussed the importance of these constituents to pain signaling, it is obvious that increasing their amounts will greatly contribute to the sensitization of the pain pathway. These are phenotypic changes that we know can potentially alter the properties of the neuron indefinitely. Under normal conditions of healing, these effects will diminish when NGF signaling ceases and the downstream components are degraded. Especially relevant is the increased synthesis of voltage-gated sodium channels because we know how essential these channels are for transmitting pain information. Moreover, studies have shown that this increase can last for months. NGF is clearly in a strong position as a target to alleviate pain.

There are as yet some unanswered questions as to how NGF can have such a plethora of effects. Since the binding of NGF to its receptor at the terminal initiates changes associated with peripheral sensitization, what distinguishes these outcomes from the activation of retrograde signaling described above? Another unresolved issue is how NGF gets out of the vesicle and into the nucleus? Nevertheless, we do not need to understand all the details, only to recognize that NGF makes a very important contribution to the development of chronic pain.

7

THE SOURCES OF PAIN

NEUROPATHIC AND CENTRAL PAIN

We have learned that the duration of the pain perceived from minor cuts, piercings, or burns is maintained by the phases of long-term potentiation (LTP) in the range of hours to days. We also know that the pain from more serious lesions can persist for much longer due to the presence of retrogradely transported injury signals such as NGF and PKG. We will now consider the response to perhaps the most serious type of injury, one that severely damages a peripheral nerve. The damage can be due to a loss of blood supply (ischemia), a localized inflammation, chemotherapeutic agents, or the transection (severing) of a nerve. Pain that originates from nerve damage is called *neuropathic pain*. In cases where a nerve is cut, all of the afferent and efferent axons within the nerve are severed, and there is a complete loss of sensation and a motor paralysis in the areas formerly served by the nerve. These are devastating injuries and there have been many efforts to restore function by bridging the gap between the cut ends of the nerve, unfortunately without much success. In addition to the loss of function, pain often develops at the cut end because the peripheral axons are still attached to the cell body.

But what is the origin of the pain? The answer might appear paradoxical because it is hard to fathom how pain can arise if the nerves are not connected to skin and muscle. To begin to answer the question, let's examine the response of a single first-order nociceptive neuron when its peripheral process is cut (see fig. 7.1).

For a very brief time after the cut, the inside of the axon is exposed to the external environment and Ca^{++} enters. The membrane then seals over, but the elevated level of Ca^{++} within will initiate enzyme cascades leading to the activation of kinases. As a result, many injury signals, including PKG and NGF, will be retrogradely transported to the cell body and will evoke pain via the usual pathway. Other retrogradely transported signals will promote a massive synthesis of constituents in an effort to regenerate the damaged axon. All of these newly synthesized components, including the kinases, ion channels, and receptors that would normally populate the terminal, are exported into the axon and transported to the site of the injury. When this massive wave of constituents reaches the cut end, it causes a dramatic swelling known as a *neuroma* (fig. 7.1). Since the membrane of the neuroma contains all of the components that would normally trigger the production of action potentials in response to a lesion, any pressure or adverse conditions can potentially elicit these potentials that will then be amplified by the LTH and result in an excruciating form of neuropathic pain.

To now address the question as to where the pain will be felt, let's consider a very extreme case where a person has lost an arm at the elbow. Action potentials arising from the neuromas beneath the skin at the stump will propagate along the nociceptive pathway to the thalamus and sensory cortex as usual, but the brain will perceive the pain as coming from the forearm or hand that no longer exists. Moreover, neuromas formed at the endings of the first-order neurons for touch, etc., can also be activated, and these

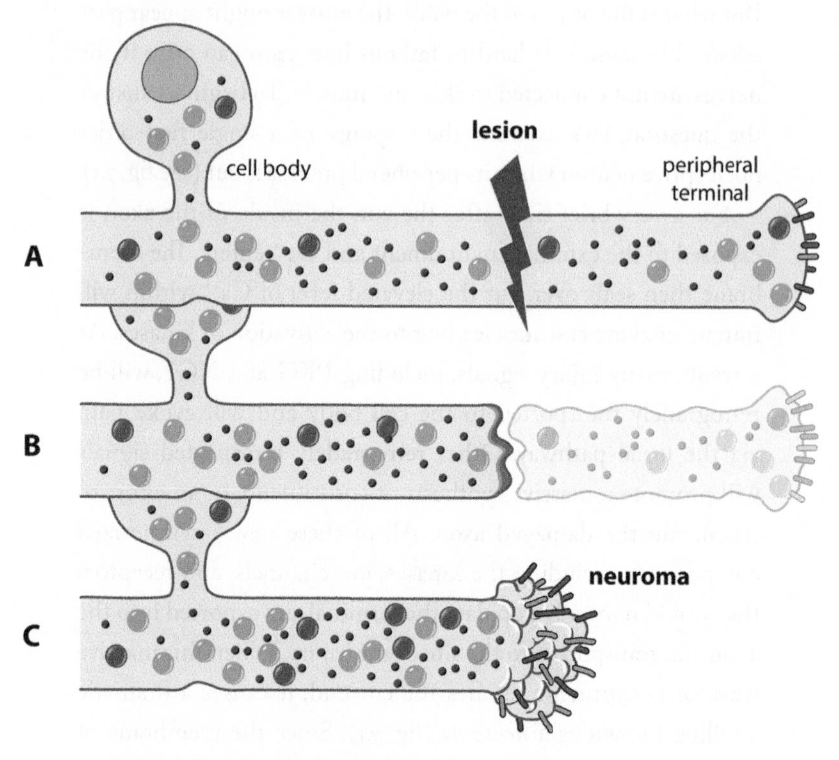

FIGURE 7.1 Formation of a neuroma. (A) A portion of the peripheral process and terminal of a first-order nociceptive neuron. The process is filled with soluble proteins (small dots) and vesicles (circles) and the terminal has receptors on its surface. (B) A lesion severs the process and the cut end of the segment attached to the cell body immediately seals over whereas the terminal segment begins to disintegrate and will eventually disappear. (C) The lesion causes an increase in the synthesis of protein and vesicles in the cell body as the neuron begins to replace the missing segment. However, there is no target, and when the newly synthesized components reach the ending, they cause it to swell, forming a neuroma. The receptors newly inserted into the surface membrane the surface of the neuroma will respond to any noxious stimuli.

potentials to the brain will also be interpreted as coming from the missing arm or hand. This is why it is not unusual for amputees to feel a ring on a finger even though the finger and arm are no longer present. These *phantom limb sensations* are among the most intractable pain conditions and even surgery to remove the neuroma usually only offers temporary relief simply because another neuroma will form. Unfortunately, phantom limb pain is an especially acute problem due to the large number of soldiers and civilians who were injured during the Iraq and Afghanistan wars.

A remarkable insight into how the brain can potentially manage pain was obtained from an ingenious approach to alleviating phantom limb sensations. The premise was that the brain misinterprets the source of the sensation because it is not aware that the limb is missing. V. S. Ramachandran at the University of California, San Diego, showed that he could alleviate the pain in some patients by using a system of mirrors so that when the patient looked at the missing right arm, for example, he saw the left, intact arm.[1] How tricking the brain works is not fully understood, but it illuminates the amazing ability of the brain to mitigate the effects of trauma. We will have much more to say about this later in the book.

What phantom limb sensations indicate is that the presence of a target is not necessary. If we think about this concept more deeply, we will realize that action potentials initiated at any point along the somatosensory pathway will result in pain, and when this happens, it can be devastating. Suppose that the second-order nociceptive neurons in the spinal cord or even worse, nociceptive circuits in the thalamus, begin to fire spontaneously. In both cases, pain will be sensed and attributed to whatever part of the sensory homunculus is activated, but of course the pain signals do not originate from any lesion in the periphery. This is called *central pain* because it arises from the activity of neurons in the central nervous system. Treating central pain is extremely challenging and is made more so because the entry of drugs to the spinal cord or brain is impeded by the presence of blood/brain barriers, and surgical interventions have a high element of risk. One innovative approach in cases where the pain originates from aberrantly active thalamic neurons has been to ablate the neurons by inserting a microprobe into the thalamus. That the ablation sometimes attenuates the pain reinforces the idea that neurons within the thalamus mediate the perception of pain.

INFLAMMATORY PAIN

We know from our discussion of the pinprick model that pain from an injury can be caused by agents released from pro-inflammatory cells of the immune system. These cells are attracted to the injury site where they are responsible for the redness, swelling, and heat that usually accompany an injury, especially one that causes extensive tissue damage. However, severe pain can also be elicited by an inflammation *in the absence of discernible physical damage*. Since an infection can be as much of a threat to survival as an injury or burn, it certainly makes sense teleologically that in addition to warding off the infection, pain would be evoked to make us aware that an infectious agent is present. We now need to consider how an inflammation generates pain.

Cytokines

An inflammatory response is a highly complex and coordinated attack that is designed to destroy pathogens but also to remove cell and tissue debris. The latter occurs normally after a rigorous workout to facilitate the repair of damaged muscle tissue. An inflammation is characterized by the recruitment of immunocytes to the area and the release of small proteins known as cytokines. There are an almost bewildering number of cytokines, but the main culprits in causing pain are the interleukins (IL) IL-1β, IL-6, and tumor necrosis factor-α (TNF-α).[2] The binding of IL-1β increases the production of substance P and the enzyme COX-2. As we already know, COX-2 synthesizes the types of prostaglandins that enhance pain. IL-6 is unique because elevated levels of this cytokine are associated with stress and the activation of centers in the brain that cause anxiety, which also exacerbates pain. We will discuss this again in chapter 12. TNF-α elevates the synthesis of itself, as well as IL-1β, and the seemingly ubiquitous NGF. The binding of TNF-α to its receptor sensitizes the area of the inflammation and the NGF increases the number of TRPV1 channels, which are then available to be activated by other inflammatory agents. The cumulative effect of all these events at terminals in the periphery is the generation of action potentials in first-order nociceptive neurons that upon reaching the synapse in the CNS will promote the release of glutamate and the activation of NMDA receptors in the second-order neurons. This should all seem familiar because we discussed these same responses in earlier chapters. It thus appears that even in the absence of an injury, cytokines can evoke the same responses and activate the same nociceptive pathway that will result in the perception of pain in the thalamus and attribution via the homunculus. A very severe immune attack can elicit a "cytokine storm" that overwhelms systems in the body; it has occurred recently in many COVID-19 cases.

Inflammations are common, and in most cases the cause of the pain is relatively easy to diagnose, such as that from arthritis, advanced diabetes mellitus, postherpetic neuropathy, or from a localized infection. Sometimes, however, the source of the pain is not at all obvious and this has long proven to be a major problem in pain management. The problem arises because the immune system is also dedicated to destroying anything "foreign." As an obvious example, consider what happens if a sponge is inadvertently left near a nerve after an operation. The sponge will be vigorously attacked by the immune system, but the cytokines that are released will infiltrate the nerve, where they will have access to the axons within. The pain occurs because the external membrane of the axons contains many of the same receptors and channels that are present at their terminals. Consequently, the binding of the cytokines to their receptors will elicit action potentials at the site of the inflammation that will propagate along the nociceptive pathway to the thalamus and cortex.[3] Moreover, if the nerve is large, many axons can be activated, and the pain will be excruciating. Clinicians call this *ectopic pain* because it does not originate at terminals in the periphery but is caused by the activation of receptors along the axon.[4] But where will the pain be felt? The answer is that it will be perceived as coming from the targets of the axons, much like phantom limb pain. A good example is the relatively common situation in which an intervertebral disc ruptures, releasing its contents onto the nearby nerves. Since the contents are normally sequestered from the immune system, they will be attacked, thereby causing an unwanted activation of the axons in the nerves. If these axons innervate the lower limb, as in sciatica, the pain will not be reported as coming from the site of the rupture, just outside the vertebral column, but rather from the thigh, lower leg or even the ankle. Actually, many components found in our bodies are considered foreign if they are found where they don't belong. Blood cells released from a ruptured spleen, or secretions from a diseased gland that enter interstitial spaces will engender an immune response accompanied by pain if there is a nerve nearby. These situations can occur anywhere in the body. We can now appreciate how difficult it can be to determine the source of the pain.

VISCERAL PAIN

Our discussions so far have focused on the pathways that convey information in response to an injury or inflammation that affects the soma, i.e., the part of the body exclusive of the viscera. Diagnosing the origin of pain is even more difficult when dealing with pain from the heart, lungs, digestive organs, glands, etc. Remember that Penfield and Rasmussen defined the homunculus, the somatosensory map along the postcentral gyrus of each cerebral hemisphere. Interestingly, the map does not have a representation of the viscera.[5] Since the homunculus reflects our ability to perceive sensations, their absence from the map indicates that the brain does not have a way to be aware of our internal organs. This of course goes against common sense. No one can dispute the pain of appendicitis or the movement of a kidney stone. There is indeed a pathway for feeling those sorts of pain, and it is important because several types of chronic pain are associated with our internal organs. Let's take a moment to learn a few general concepts about how our viscera are innervated.

THE TWO-WORLD VIEW OF NERVOUS SYSTEM FUNCTION

Our nervous system is designed so that the brain receives information about events in the outside world via afferent neurons associated with vision, touch, pain, etc. We also know that the brain evaluates this information and responds via the motor neurons that activate the appropriate muscles. However, information from the inner world that consists of the heart, lungs, liver, kidneys, and digestive system is conveyed by a separate, visceral nervous system. It is composed of two afferent components that are dedicated to assessing the function of our visceral organs. The first comprises neurons that send signals from the viscera to localized centers lower in the brain. These signals provide information regarding the status of our organs so that at every moment in time, and without our conscious awareness, our heart rate, blood flow, and other essential functions are monitored by these neurons. We are not directly aware of this information simply because the brain does not have circuits capable of processing this information into a sensation. This arrangement maximizes efficiency and can adjust rapidly to conditions. There is some debate as to whether we can be indirectly aware of the workings of our inner organs, and some evidence indicates that our mood can be influenced by information conveyed via these visceral afferents. Hence, these neurons comprise an *enteroceptive* system.

The response to the enteroceptive inputs occurs via the *autonomic nervous system*, which consists solely of motor neurons. Depending on the input, the motor neurons will speed up or slow down the heartbeat, increase or decrease the movement of digested food along the digestive tract, or augment responses by activating glands to release hormones and/or neurotransmitters. This ingenious design means the brain does not need to use valuable circuitry to take care of routine housekeeping functions.[6]

The second afferent component of the visceral nervous system, which is of greater relevance to us, is comprised of first-order nociceptive neurons that send signals from the viscera when something is wrong. These signals are translated into pain by an indirect process that will be discussed below. Thus, we actually have two different dedicated nervous systems: a somatic system that deals with the world external to our bodies and a visceral system that regulates the workings of our internal organs and alerts us when an organ is threatened. Within these two independent systems is a boundary that separates their territories and also separates the pathway responsible for visceral pain from that responsible for somatic pain. The viscera are covered by a visceral membrane that is in direct contact with the surface of each organ, and a parietal membrane that lines the cavities that contain the organs and is in direct contact with the inner surface of the soma. Nociceptive information from the parietal membranes is conveyed by branches of the spinal nerves, whereas information from structures enclosed within the visceral membranes is conveyed by the visceral nerves.

VISCERAL PAIN IS REFERRED

We already know how pain from somatic structures is communicated to the brain; now we need to describe how the brain receives information about a lesion from the viscera. Surprisingly, except for hunger, the primary sensation that is perceived from the viscera is pain. Touching, cutting, or otherwise manipulating the internal organs does not elicit a response. Moreover, the pain that arises from the viscera is elicited only by an inflammation or in response to an expansion, such as a kidney stone passing along the ureter. Unfortunately, most forms of cancer do not cause these types of disruption and can therefore develop without pain. It is quite amazing in anatomy labs to see cadavers from donors who were seventy-five or even eighty years old whose internal cavities are filled with tumors that must have grown over many years without symptoms.

Pain from the viscera is communicated by a subset of first-order nociceptive neurons whose cell bodies reside within the dorsal root ganglia along with all the first-order nociceptive neurons that innervate the body. The peripheral terminals of these visceral neurons are located on their target organ and the peripheral processes initially course through the visceral nerves that contain the motor axons of the autonomic nervous system (fig. 7.2). Each visceral nerve joins a ventral primary ramus of a spinal nerve. When the peripheral processes of the first-order visceral nociceptive neurons reach this junction, they continue within

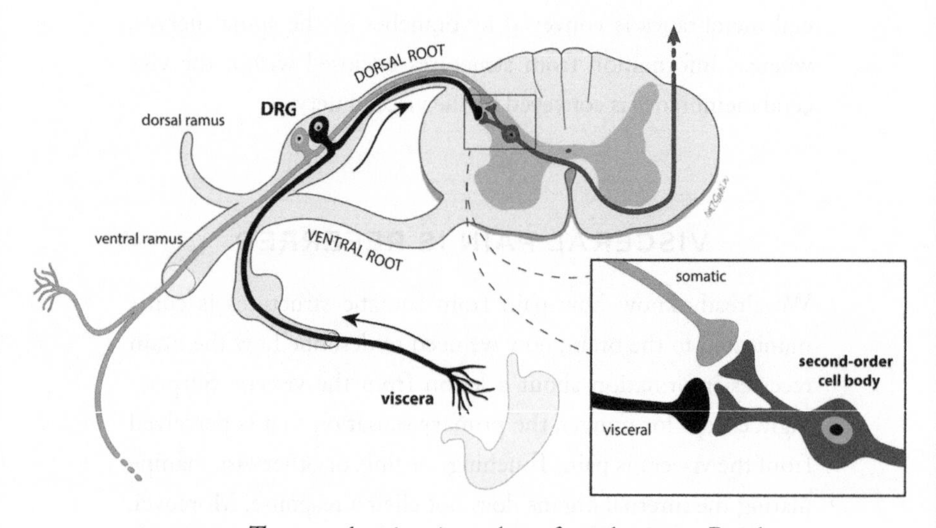

FIGURE 7.2 The general nociceptive pathway from the viscera. Peripheral axons of first-order neurons from the viscera course within a visceral (autonomic) nerve to the branch of a spinal nerve where they accompany peripheral first-order nociceptive axons coming from somatic targets. Both sets of axons pass through the dorsal root and their central processes synapse on second-order neurons. According to current theory, both inputs are onto the same group of second-order neurons (inset). Consequently, when signals from the second-order neurons ascend to the thalamus and sensory cortex, they are interpreted by the brain as originating from the somatic target, not the viscera. Thus, the pain is referred.

the spinal nerve, pass through the dorsal root, and their central process synapses on the second-order neurons in the dorsal region of the spinal cord. Note that the central process of the visceral nociceptive neurons enters the spinal cord at the same level as the central processes of the first-order nociceptive neurons that innervate the skin, parietal membranes, or other target in the periphery. What happens next is not certain, but an accepted explanation is that the central processes of both the first-order somatic and first-order visceral neurons synapse on the same second-order neurons (fig. 7.2). The signals from the second-order neurons ascend to the thalamus, which communicates with the homunculus. However, since axon potentials from second-order neurons at this level of the spinal cord usually respond to a lesion to the skin, the brain misinterprets these signals and assigns the pain to the somatic target.[7] In other words, pain that should be perceived as coming from the heart or some other organ is instead referred to the body at the level of entry of the spinal nerve.[8]

This seems rather bizarre, and it certainly complicates efforts to diagnose the source of the pain. Fortunately, there are maps that show the area of the soma to which pain from the various organs originates (fig. 7.3). To better understand this from a clinical perspective, we can use appendicitis as an example. The appendix is located in the lower right quadrant of the abdomen and is innervated by visceral nerves that join the tenth thoracic spinal nerve. When the appendix becomes inflamed and swells, action potentials are elicited that propagate within the visceral nerve to the tenth thoracic spinal nerve and then to the synapse on the second-order neurons at the T10 level of the spinal cord. The first-order nociceptive neurons in the tenth spinal nerve, which activate the same second-order neurons, innervate structures all along the tenth dermatome, including the umbilicus

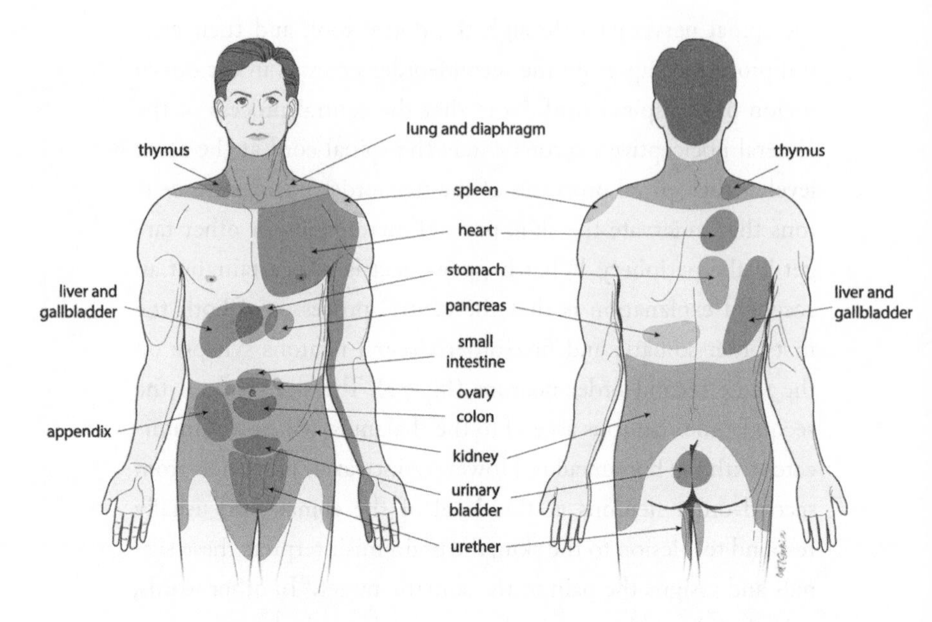

FIGURE 7.3 Map showing approximate location of pain referred from a lesion to the viscera.

(belly button). Consequently, pain from the initial stages of appendicitis is perceived as coming from the umbilical region.

Now, let's consider what happens if the inflamed appendix bursts. Material released from the appendix will generate an inflammation on the parietal membrane above the appendix. This will activate the peripheral processes of the first-order nociceptive neurons within spinal nerves that supply the area, and the result will be pain that is perceived as coming from the skin directly above the appendix. The shift in the perceived origin of the pain is one characteristic of appendicitis.

The idea that the pain is referred because both the somatic and visceral nociceptive neurons synapse on the same second-order neurons might be reasonable, but it is incomplete because it does not account for the fact that visceral and somatic pain differ in two important respects. Thus, the initial pain from the appendicitis is visceral and will be felt as dull and diffuse. When the appendix bursts, however, the pain is somatic and is sharp and localized, as would be any pain from the skin.

Even more paradoxical is what happens when the spleen ruptures. The spleen is located in the left upper region of the abdominal cavity, just below the diaphragm, so where will the patient feel the pain? The answer is in the left shoulder (fig. 7.3). The blood cells from the rupture leak into the adjacent space where they do not belong and where they are attacked by the immune system. The pro-inflammatory cytokines that are released diffuse onto the diaphragm and initiate action potentials in its phrenic nerve. The phrenic nerve enters the spinal cord at the same level as the somatic nerve that innervates the shoulder. Consequently, when the action potentials from the phrenic nerve activate the second-order neurons, the brain mistakenly attributes the pain to the shoulder.

We can now appreciate how difficult it can be to determine the source of pain because it can be caused by an injury or inflammation, can be neuropathic or central, and can be referred from the viscera. We will learn in chapter 10 it can also have a psychological source. As we said in the introduction, pain is complex!

Except for central pain, it should be evident from all that has been presented so far that neuroscientists have identified many of the molecular components in the nociceptive pathway that are essential for the transmission of pain information. While this understanding is a great accomplishment, it is only part of the story; more recent studies have shown that this pathway and pain are modulated by external neuronal circuits that originate higher in the brain. We will describe these circuits in the next chapter and show how they have greatly widened our understanding of how the perception of pain is influenced by external events.

II

THE MODULATION OF PAIN BY CIRCUITS IN THE BRAIN

8

THE EXTERNAL MODULATION OF PAIN

Descending Systems

INTRODUCTION: A NEW PERSPECTIVE

The previous chapters discussed the somatosensory system that consists of the first- and second-order neurons in the nociceptive pathway and the third-order neurons in the thalamus that project to the sensory homunculus in the postcentral gyrus of the cerebral cortex. This system provides information about an injury or lesion. For many years, efforts to alleviate pain have focused on preventing this information from being transmitted along its pathways. While the molecular events that unfold in the somatosensory system adequately describe the initial response to a typical injury or inflammation, recent advances in neuroscience and psychology have forced us to reevaluate what it means to be in pain, i.e., to hurt. We have been taught, and learn through experience, that pain is just another sensation like touch or vision, but we now know that this is not correct. In fact, the somatosensory system is only one component of an extensive neuronal network in the brain. This network ensures that what we ultimately experience as hurtful is shaped by past experience, mood, and present circumstances. How this occurs will be the subject of several subsequent chapters. Identifying and characterizing the circuits in the brain that regulate what we perceive as pain has major implications for managing chronic pain, and we will begin by presenting a very dramatic example of how the brain can control pain.

PAIN, CIRCUMSTANCE, AND OPIATES

Stress-induced analgesia was a well-documented phenomenon during World War I, when soldiers with grievous wounds disregarded their pain in order to escape danger. Actually, they were not disregarding the pain; they were not aware of the pain. Awareness is one of the complexities that we will have to confront later. Nevertheless, it was clear from the experiences of the soldiers that pain is not an automatic response to an injury. Stress-induced analgesia was originally thought to occur only when ignoring an injury is necessary for survival, and while this makes sense, survival is not the only reason. There have been many reports of workers suffering serious injuries in industrial accidents where they don't remember experiencing any immediate pain. One man sawed through three of his fingers and didn't realize what he had done until he saw the blood. We mentioned earlier that synapses are sites of regulation, and we will now learn that all these effects are due to extrinsic pathways that alter the efficacy of the synapses between the first- and second-order neurons.

The Search for Endogenous Opiates

One explanation for stress-induced analgesia was predicated on the idea that the body has an internal mechanism for dealing with pain. A major proponent of this hypothesis was Professor

Hans Kosterlitz at the University of Aberdeen, Scotland, in the 1960s. He was well aware of the wartime reports, and he put that information together with the known analgesic properties of morphine to postulate that the opiate was mimicking a substance made in the body that could blunt pain. He called this hypothetical endogenous compound an "endorphin," which was shorthand for "in the body morphine." This simple but logical connection launched one of the most fascinating sagas in the history of pain management.[1]

Proposing that an endorphin exists was a novel and intriguing idea: proving its existence was far more difficult and required two very different approaches. The first was to try to isolate the putative endorphin and the members of Kosterlitz's lab were fortunate in correctly assuming that the compound would be present in the brain. They were also fortunate because they had an assay that could be used to screen for the compound. Morphine causes constipation by blocking peristalsis in the intestines and this could be demonstrated directly in the lab using an isolated guinea pig ilium, which is the terminal region of the small intestine. The idea was to elicit peristalsis, add the test compound, and see if it inhibited peristaltic contractions in a *reversible* way. Reversibility was important because it was possible that many compounds would damage or kill the muscles responsible for the movement. The best source of the potential endorphin was the pig brain because it could be obtained in large quantities from a local abattoir. The protocol was to homogenize the brains, divide the homogenate into fractions based on certain criteria, and then test each fraction using the ileum peristalsis assay. This arduous task was assigned to John Hughes, and that it eventually proved successful was a tribute to his perseverance. Luck was surely involved because the compound could exist, but be broken down during the extraction procedure.[2] He found that a soluble fraction mimicked the effects of morphine on the ileum, but it contained many proteins and other compounds and isolating the specific factor was proving difficult.

Meanwhile, groups in the United States were studying the molecular composition of synapses. Not much was known about these structures back in the 1960s and early 1970s. Nevertheless, Kosterlitz and others predicted that the endorphin would block pain by modulating synaptic function: a simple idea with profound consequences. We have already discussed how neurotransmitters initiate an outcome by binding to highly specific receptors embedded in the membrane of their target. If Kosterlitz and his colleagues were correct, then synapses should have a receptor that recognizes the endorphin. As it turns out, this was a propitious time to explore this possibility and several labs around the world were well situated to find these receptors. The structure of morphine was known, as were a number of derivatives, including naloxone, a very potent opiate antagonist. Naloxone binds to the putative receptor even more tightly than the endogenous opioid, and this connection is important because it can be used to verify that a response is actually due to the opioid.[3]

Important also were the newly developed processes for synthesizing compounds labeled with tritium (^{3}H), a radioactive isotope of hydrogen. Thus, binding could be detected by following the radioactivity, thereby obviating the need to do thousands of laborious assays. In 1973, Candace Pert and Solomon Snyder at Johns Hopkins added ^{3}H-naloxone to a homogenate of brain tissue. The homogenate was then separated into fractions, and by following the distribution of the ^{3}H-naloxone, they identified a morphine receptor, which was abbreviated μ (Mu, for morphine), in the membrane fraction. This outcome is generally considered to be the first identification of an endorphin receptor in the brain, although it was actually the third to be found, after the delta (δ) receptor in the vas deferens of mice, and the κ (kappa) receptor that was characterized by its pharmacological properties.[4] The kappa and delta receptors were subsequently found in the brain, which dramatically changed the thinking about a single endorphin because it was now likely that there would be more than one. This theory was confirmed in 1975 when Kosterlitz and his colleagues published their finding that the brain contained two endorphin-like compounds that they called *enkephalins* (ENK). They were not proteins, however, but pentapeptides comprised of five amino acids: Met-enkephalin (Tyr-Gly-Gly-Phe-Met) and Leu-enkephalin (Tyr-Gly-Gly-Phe-Leu). The finding of the enkephalins in the brain was a milestone in the understanding of nociception. The enkephalins were not alone, however, and other studies soon showed that they were only one of three classes of molecules with endorphin-like properties. The other two are the endorphins proper and the dynorphins. Thus, the search for an endorphin wound up finding three classes of endorphins and three receptors. The next challenge was to determine where in the brain the receptors were located.

Distribution of the Opioid Receptors

Once again ^{3}H-labeled ligands were put to good use by adapting a new procedure known as radioautography. Basically, the distribution of an opioid receptor could be determined by first exposing a thin section of brain tissue to a radioactive ligand, say ^{3}H-naloxone, that would bind strongly and with specificity to the receptor in the tissue. The tissue section was then coated with an emulsion that turns black in the presence of radioactivity, much as film turns black in response to light. After an incubation period, areas that contain the ^{3}H-ligand bound to the receptor are blackened and can be easily distinguished from surrounding areas. The results were quite striking in that Mu opioid receptors were found in neurons in the dorsal region of the spinal cord, which we know is the location of the synapse between the first- and second-order nociceptive neurons, and in a group of neurons in the midbrain that comprise a structure known as the periaqueductal gray (PAG).[5] This latter finding was especially significant because some studies showed that electrical stimulation of the PAG could diminish pain without interfering with touch, pressure, or temperature sensations. In fact, the antinociceptive effect of the stimulation was so powerful that it was possible to perform surgery on a fully conscious rat without causing distress to the animal. Finally, subsequent experiments demonstrated that the analgesia could be prevented by injecting naloxone, the Mu receptor antagonist.

Taken together, these results provided compelling evidence that the endogenous endorphins and their receptors comprise an inherent system that is responsible for shutting down pain under conditions of severe trauma. They also placed the PAG as an essential center for the perception of pain.[6]

Viewed from a neuroanatomical perspective, the most parsimonious interpretation of the data is that some axons of second-order nociceptive neurons in the spinal cord form an ascending pathway that terminates on the neurons in the PAG (fig. 8.1). This pathway is activated only after a severe or traumatic injury that elicits a barrage of action potentials. These potentials propagate along this pathway to stimulate the neurons in the PAG that contain the enkephalins. The action potentials elicited in these neurons then descend along their axons to the dorsal horn of the spinal cord where they promote the release of the enkephalins, thereby blocking synaptic transmission between the first- and second-order neurons. In other words, the stress-induced

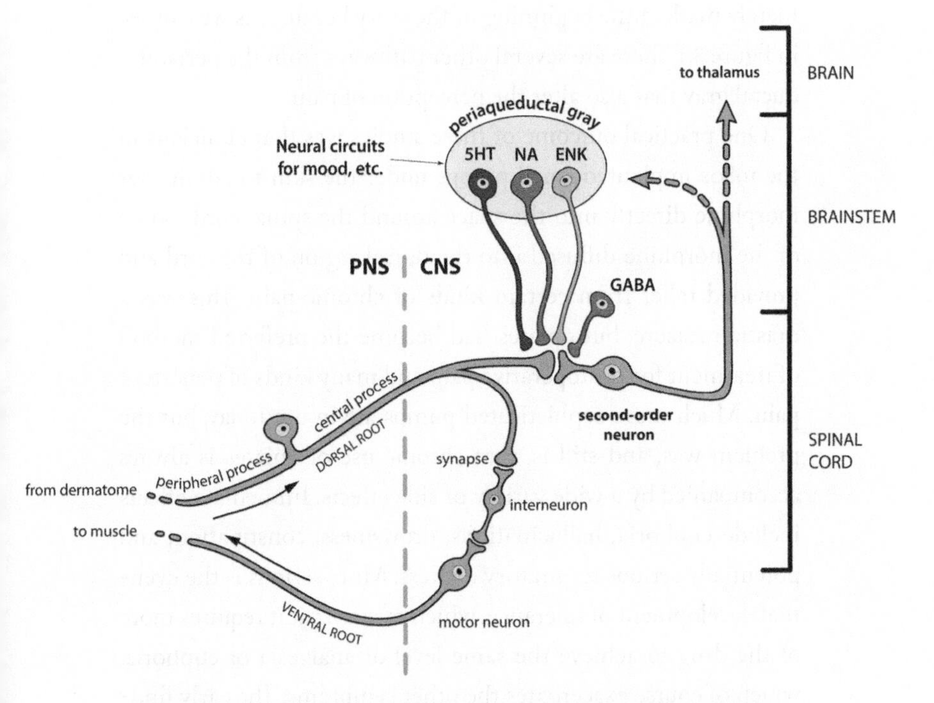

FIGURE 8.1 Descending pathways from the periaqueductal gray to the synapse between the first- and second-order nociceptive neurons in the spinal cord. The release of an encephalin (ENK) results in analgesia by blocking the transmission of nociceptive signals. The enkephalin neurons are activated via an ascending pathway depicted as a branch from axons of second-order neurons to the thalamus (dotted lines). Descending pathways from neurons containing serotonin (5HT), noradrenalin (NA), and gamma-aminobutyric acid (GABA) also modulate the synapse. The pathway for 5HT is not direct. Note that the periaqueductal gray also receives inputs from neural circuits in the brain.

analgesia experienced after a grievous injury occurs because the trauma induces the release of enkephalins that block the nociceptive signals from the periphery (fig. 8.1). Quite a revelation! Moreover, it was the first indication as to how neurons in the brain could influence the perception of pain. However, this merely marked the beginning of the story because, as we can see in figure 8.1, there are several other pathways from the periaqueductal gray that also alter the perception of pain.

One practical outcome of these studies was that clinicians in the 1980s implanted small pumps under the skin to administer morphine directly into the space around the spinal cord. Some of the morphine diffused into the dorsal region of the cord and provided relief from certain kinds of chronic pain. This was a drastic measure, but opiates had become the preferred method of treatment for postoperative pain and many kinds of persistent pain. Much more sophisticated pumps are in use today, but the problem was, and still is, that chronic use of opiates is always accompanied by a wide variety of side effects. Immediate effects include euphoria, hallucinations, drowsiness, constipation, and potentially serious respiratory distress. More serious is the eventual development of tolerance, which means that it requires more of the drug to achieve the same level of analgesia or euphoria, which of course exacerbates the other symptoms. The early findings that opiate receptors were present in the vas deferens indicated that they would not be restricted to the PAG and spinal cord. Indeed, radioautographic studies soon revealed that opioid receptors are widely distributed throughout the brain. Disruption of these systems by chronic opiate abuse perfectly explains the physical manifestations of addiction that we mentioned.

MECHANISM OF OPIOD ACTION IN THE SPINAL CORD

To understand how the release of the enkephalins in the spinal cord attenuates pain, we need to examine the location and function of the opioid receptors. The general structure of these receptors is similar to that of other receptors we have discussed and consists of an extracellular region that contains the binding site, seven transmembrane helical loops, and an intracellular terminus. Most important is that the Mu receptors are located on the *presynaptic terminal* of the first-order nociceptive neurons (fig. 8.2).

When the action potentials propagating along axons from the periaqueductal gray reach the synapse on the presynaptic terminal, the enkephalins are released into the cleft. They bind to

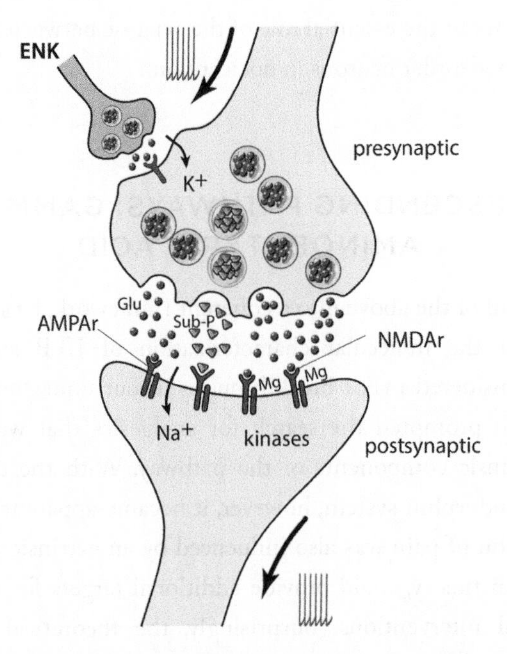

FIGURE 8.2 Axons of the enkephalin (ENK) neurons terminate in a synapse with the presynaptic terminal of the first-order nociceptive neurons. When the ENK neurons in the periaqueductal gray are activated in response to a traumatic lesion, the ENK is released. It binds to its receptor resulting in an influx of K^{+} that blocks the release of glutamate and substance-P and prevents the activation of the second-order neurons.

their recognition site on the Mu receptor, eliciting a conformational change in the intracellular part of the receptor in the presynaptic terminal that activates kinases with two primary effects. First, they open potassium channels, which hyperpolarizes the terminal and makes it resistant to the action potentials coming from the periphery. Second, they block the release of the calcium ions into the terminal, which we know is necessary for the mobilization of synaptic vesicles. The net result is no release of glutamate or substance-P, no activation of either the AMPA or NMDA receptors on the postsynaptic membrane, and a strong analgesic effect. All of this makes perfect sense from what we learned about the essential role of the synapse between the first- and second-order neurons in nociception.

DESCENDING PATHWAYS: GAMMA-AMINOBUTYRIC ACID

To put all of the above into a coherent framework, let's remember that the molecular characterization of LTP and LTH were considered major breakthroughs in our understanding of pain and prompted the search for analgesics that would target intrinsic components of the pathway. With the discovery of the endorphin system, however, it became apparent that the perception of pain was also influenced by an extrinsic pathway, which, in theory, could provide additional targets for pharmacological interventions. Surprisingly, the theoretical became practical rather quickly when it was discovered that antianxiety drugs such as Valium had antinociceptive properties. Here then was evidence that the encephalin pathway was not alone: there were other extrinsic neurotransmitter pathways that influenced pain. Moreover, these pathways were not part of the nociceptive system but were linked to mood and higher brain functions (figs. 8.1 and 8.2).

Valium targets pathways that use gamma-aminobutyric acid (GABA) as their neurotransmitter. GABA is widely distributed in the brain and spinal cord, but its function is distinctly different from other neurotransmitters because it doesn't excite a postsynaptic target; rather, it reduces or prevents synaptic transmission.[7] The notion that the brain contains an inhibitory neurotransmitter first seemed strange until it was discovered that one of GABA's functions is to prevent synapses from firing excessively. Thus, when GABA activity is diminished, neurons in the brain fire more than they should, which can lead to anxiety, stress, an increased heart rate, high blood pressure, and a host of other problems. Anxiety in particular enhances pain. Obviously maintaining an optimal level of GABA functioning is a challenge for clinicians and the pharmaceutical industry.

GABAergic neurons are found in regions of the dorsal horn that we know are important for transmitting pain impulses. As we might predict, the axons of these neurons synapse on the presynaptic terminal of the first-order nociceptive neurons (fig. 8.3). There are two types of GABA receptor, A and B, but it is the A type that is embedded in the membrane of the terminal.[8] It has five subunits, each of which has several transmembrane regions and a very long external segment that contains recognition sites for GABA. The subunits can join in various combinations to form the functional receptor with a central canal that regulates the entry of chloride (Cl^-) ions. Note the negative charge: when GABA binds to its recognition site on its receptor, the channel opens and the Cl^- enters and hyperpolarizes the terminal by making it more negative. This reduces the probability that action potentials arriving from the site of a lesion will be able to depolarize the terminal to cause the release of glutamate.

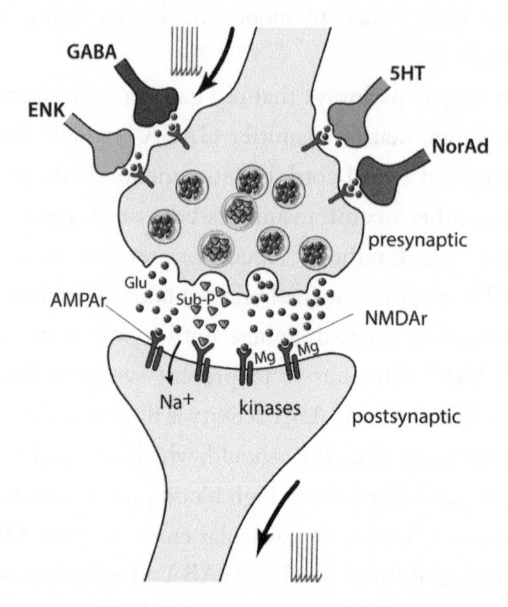

FIGURE 8.3 We have already discussed the function of the ENK system. Here we see other descending pathways that modulate pain. Axons from neurons that contain serotonin(5HT), noradrenalin, (NorA), or gamma-aminobutyric acid (GABA) form synapses with the presynaptic terminal of the first-order nociceptive neurons. Stimulation of these neurons results in the release of their neurotransmitter, thereby modulating the release of glutamate and substance-P and influencing the communication with second-order neurons in the nociceptive pathway.

In other words, the nociceptive pathway is inhibited or modulated depending on the level of GABA. A disease or defect that reduces GABA levels in the spinal cord would allow more synaptic activity between the first- and second-order neurons and might contribute to hyperalgesia. Drugs like pregabalin (Lyrica) are important because they increase the levels of GABA, thereby enhancing its effects.[9] However, we must remember that the function of GABA and its receptor is to prevent excessive synaptic activity. Consequently, it is not clear whether artificially elevating GABA levels will significantly block synaptic transmission in response to a lesion. Lyrica has been approved by the FDA to treat only some types of pain; it has proven difficult to design truly effective GABA analgesics for the routine management of pain. The main impediments are side effects such as dizziness and sedation due to the activation of GABA receptors in the brain, most of which are not involved in the perception of pain.

DESCENDING PATHWAYS: SEROTONIN AND NORADRENALIN

One would imagine that having both opioid and GABA pathways as extrinsic modulators of the nociceptive pathway would be sufficient to account for all the ways in which the perception of pain can be influenced by external circumstances.

However, nature likes redundancy, especially for essential functions, and there are two additional descending pathways that we need to discuss because they are important targets for the pharmacological management of pain. One consists of neurons that contain the neurotransmitter serotonin (also called 5-hydroxytryptamine or 5HT) and the other, noradrenalin (or norepinephrine). We know that these neurotransmitters are important because clinicians have found that tricyclic antidepressants that target these pathways can mitigate certain kinds of pain.

The cell bodies of both the noradrenergic and serotonergic neurons reside in centers located in the brain, and studies have shown that these centers receive inputs from circuits responsible for fear, anxiety, and other emotions. Noradrenergic neurons in particular are associated with the ability to sustain attention, and this will be important in a subsequent chapter. Axons from both the serotonergic and noradrenergic neurons descend and form presynaptic terminals on the presynaptic terminal of the first-order nociceptive neurons (fig. 8.3). The antinociceptive action of the noradrenergic pathway occurs when action potentials arriving at the presynaptic terminal cause the release of noradrenalin into the synaptic cleft. The noradrenalin binds to alpha-adrenergic receptors in the membrane of the presynaptic terminal, resulting in the activation of agents that inhibit the presynaptic voltage-gated Ca^{2+} channels. Since Ca++ is necessary to mobilize synaptic vesicles for the release of glutamate, this prevents the activation of the second-order neurons and no signals are sent to the thalamus. The role of serotonin is more complex because there are at least twelve different 5HT receptors that can influence pain processing. What seems to be certain is that the release of serotonin in the dorsal horn also diminishes communication between the first- and second-order nociceptive neurons.

ENHANCING THE LEVELS OF NORADRENALIN AND 5HT

These descriptions provide a basic framework for understanding how these two neurotransmitters can have antinociceptive affects. The reality is more complex because both the serotonergic and noradrenergic pathways also synapse on other neurons in the dorsal horn that might block neurotransmission in other ways.

Nevertheless, the one important fact to remember is that regardless of these complexities, the tricyclic antidepressants that influence these pathways have proven useful in ameliorating certain kinds of pain.

These drugs function in two ways. First, we know from experience that protracted pain causes anxiety, which can lead to a progressive depressive state accompanied by enhanced pain sensations. Thus, like the GABA agonists, the antidepressants can mitigate pain by relieving the depression. However, it can take weeks or months before the depression is relieved, whereas certain antidepressants can relieve pain within days. The discrepancy in timing suggested that there was an analgesic effect that differed from the effects on depression, and this prompted efforts to characterize the mechanisms responsible for the analgesia.

The efficacy of a neurotransmitter is governed by how long it is present in the cleft at a high enough concentration to activate its receptor. Since there was good evidence that elevated levels of some neurotransmitters had beneficial effects, it was important to determine what processes regulate the fate of neurotransmitters in the cleft. Some are simply degraded by a specific enzyme, but there are two much more clinically important processes that reduce the levels of 5HT or noradrenalin. The first involves their removal from the cleft by systems that transport each neurotransmitter back into the presynaptic terminal. The tricyclic antidepressants block this uptake mechanism, thereby maintaining an effective level of 5HT or noradrenalin in the cleft for a longer period. A somewhat surprising finding was that the uptake is also blocked by ingredients found in marijuana (cannabis); we will have more to say about this in the next chapter.

The second process involves the fate of the neurotransmitter once it has been taken up. Some of it is repackaged into vesicles for release and some is destroyed by enzymes known as monoamine oxidases. Thus, the drugs commonly known as monoamine oxidase (MAO) inhibitors prevent the breakdown of serotonin and noradrenalin so that more is available for rerelease into the cleft. MAO inhibitors are useful at relieving symptoms associated with depression, such as sadness or anxiety, but were also found to be effective for some types of chronic pain. As we would expect, however, they have serious side effects, including a withdrawal syndrome on discontinuation, and are now used only when other antidepressants have proven ineffective.

Although the tricyclic antidepressants can relieve some pain, a breakthrough of sorts was the finding that the serotonin transporter was different than that for noradrenalin, which led to the development of drugs that selectively prevented the uptake of serotonin.[10] These selective serotonin reuptake inhibitors (SSRIs)—including Prozac, Lexapro, Paxil, and Zoloft—were highly successful as mood elevators and were soon followed by norepinephrine reuptake inhibitors (NRIs) such as Reboxetine. The ultimate breakthrough, however, was the development of serotonin and noradrenalin reuptake inhibitors (SNRIs). Duloxetine was the first of the dual inhibitors to be approved by the USFDA for treatment of pain from diabetic neuropathy, and a review of the various uptake analgesics have shown that the SNRIs have the best analgesic efficacy against pain from fibromyalgia and osteoarthritis. Since none of these drugs are universally effective in alleviating pain, however, they are broadly considered as adjuncts to be used in conjunction with other treatments. Unfortunately, the further development of these uptake analgesics has been hindered by the fact that the responsiveness of the receptor systems appears to vary with the type and duration of the pain and the mode of drug administration. In addition, these drugs have only a limited effectiveness in treating other types of chronic pain.

By now it should be apparent that our claim that pain is the most complicated sensation is not an exaggeration. In previous chapters, we discussed the importance of the somatosensory system in processing information from a lesion. What we have now learned is that this system is only part of the story because it does not function in isolation; it is significantly influenced by the pathways that can modulate pain by releasing the endorphins, GABA, etc. These neurotransmitters fine-tune the transmission of pain impulses across the synapse between the first- and second-order neurons. Moreover, they do not function at the same time, but each relieves pain from a specific source. Thus, the endorphin system blocks pain after grievous injuries, GABA prevents over-activation in response to normal injuries, and serotonin and noradrenalin reduce pain in order to "elevate" mood. The elevation is highly significant because it links the control of pain to centers in the brain that regulate mood, anxiety, and attention. Not only does this begin to explain why pain is subjective but, as we will see in subsequent chapters, the identification of the neuronal networks in the brain responsible for these inputs has resulted in several non-pharmacological approaches to managing pain. This does not diminish the importance of developing new analgesics, and in the next chapter we will discuss several promising targets and describe the scientific and bureaucratic hurdles that must be cleared to get an analgesic drug to the marketplace.

9

ALLEVIATING PAIN

The Pharmacological Approach

DRUG DEVELOPMENT

The successful development of a potent, selective analgesic takes years, is very expensive, and is fraught with difficulties and challenges. The process requires multiple steps, and stringent criteria determine which potential drug candidate progresses to the next level. Very few compounds that begin the developmental process actually become drugs that are available to the populace. We will first discuss several approaches to identifying a potential target for drug development and will then describe the magnitude of the problems that must be overcome as a way of explaining why it so difficult and expensive to bring a drug for chronic pain to market.

TARGET SELECTION

Opium and Willow Bark

The crucial step in developing an analgesic is to first identify a target molecule that has promise for alleviating pain. The earliest discoveries took advantage of what nature had to offer, and it is worth noting that many of our most effective analgesics today were actually recognized in ancient times, albeit in crude form. The ancient Greeks and Egyptians knew about the analgesic properties present in the unripe seeds of the poppy plant, *Papaver somniferum*. Concoctions and elixirs made of poppy seed extracts were widely used for centuries,[1] but their true potential was not realized until advances in chemistry resulted in the purification of the active ingredient, the opiate morphine. Morphine is appropriately named after Morpheus, the Greek god of sleep, because of its tendency to induce drowsiness. The first industrial production of morphine occurred in 1825 by a firm in Germany; it would grow to become Merck, the giant pharmaceutical company. Once the effective agent was identified, the next step was to have medicinal chemists synthesize derivatives in the search for even more effective forms of the drug. From morphine came codeine, oxycodone, and fentanyl, and from cocaine came procaine, lidocaine and a host of other analgesics.[2]

An alternative to opium in some ancient cultures was an extract of willow bark, and in 1820, the active ingredient was identified as salicylic acid. Forty years later salicylic acid was mass-produced, but it proved disappointing because the pure form caused diarrhea and vomiting. A search for alternatives ensued, and in the late 1890s, Felix Hoffman of the Bayer Company in Germany found that acetylsalicylic acid was a better choice. It was marketed under the trade name Aspirin, which became the most widely used analgesia in the world.

Thus, two giant pharmaceutical companies were founded, at least in part, by taking advantage of ancient remedies for pain.

The approach described above was based on empirical evidence that the drugs extracted from the elixirs and potions were effective, but how the drug actually produced analgesia was not known, and this incomplete understanding hampered the ability to refine the drug to increase its efficacy and reduce side effects. Aspirin, for example, can irritate the stomach and it took a long time after aspirin was synthesized for scientists to discover that it inhibits the enzyme cyclooxygenase (COX), which we know from a previous discussion converts arachidonic acid into precursors of the prostaglandins. The latter are important mediators of inflammation and the resultant pain, fever, and the dilation of arteries. Dilation of arteries in the head produces pain. Aspirin and other COX inhibitors such as Ibuprofen and Naproxen became best-selling analgesics for headache and were collectively known as nonsteroidal anti-inflammatory drugs (NSAIDs). The NSAIDs have one major drawback: because some prostaglandins protect the gastrointestinal tract from acid, inhibiting COX can irritate the stomach lining and repeated use can cause ulcers. In 1988, scientists at Brigham Young University discovered another cyclooxygenase, COX-2. Unlike the original COX (now COX-1) that is constitutively present in cells, COX-2 is present primarily in cells involved in inflammation and its levels increase in the presence of inflammatory agents such as cytokines. Moreover, COX-1 and COX-2 produce prostaglandins with different properties. COX-1 protects the gastrointestinal tract from acid assault whereas COX-2 produces prostaglandins involved in pain, fever, and inflammation. Significantly, the NSAIDs inhibit both COX-1 and COX-2. Obviously, it would be profitable to obtain a selective inhibitor of COX-2, and this was possible because the enzymes differ in structure. Three years after its discovery, a selective COX-2 inhibitor was synthesized by the Searle group at Monsanto and was eventually marketed by Pfizer under the trade name Celebrex, which is a *selective* NSAID, whereas Aspirin and similar drugs are *nonselective* NSAIDs. By selectively blocking COX-2, Celebrex relieves pain from conditions such as arthritis without endangering the stomach. This story illustrates two points. First, development of an effective drug is made far simpler if the target is known, and this knowledge was largely responsible for the beginnings of the pharmacological industry. Second, as was the case with COX-2, many targets are identified by scientists working on basic research in university laboratories. Since the funding for this work comes largely from the National Institutes of Health or the National Science Foundation, which are government agencies, much of the essential work for drug development is paid for not by the pharmaceutical industry but by our tax dollars.

Marijuana

Another natural source of pain relievers are the flowering plants in the genus Cannabis. The dried flowers or leaves of *Cannabis sativa*, known as marijuana, have been used for centuries for their hallucinogenic, euphoric, antispasmodic, and analgesic properties.[3] Extracts of the marijuana plant seem to be especially effective for relieving neuropathic and inflammatory pain but with a variety of undesirable side effects.[4] Attempts to identify the agents that are responsible for the analgesia were limited in the United States because marijuana was considered a substance of abuse. The restrictions were partially lifted in the 1990s and research has subsequently shown that marijuana is a promising source of novel targets for the development of drugs to treat all types of pain.

Marijuana contains over one hundred different compounds (cannabinoids), but many of its most obvious behavioral effects can be attributed to delta-9-tetra-hydro-cannabinol, commonly known as THC. THC relieves pain by binding to two receptors, CB1 and CB2. Both are typical seven-folded transmembrane proteins that mediate signal transduction. We have discussed similar receptors in previous chapters. The CB1 receptor is widely distributed throughout the nervous system whereas the CB2 receptor is located primarily in the periphery. Most significant, however, is that the CB1 receptor has been found on neurons in the periaqueductal gray (PAG), dorsal root ganglia, and in the dorsal regions of the spinal cord that receive input from first-order nociceptive neurons. Lower levels are found in the thalamus. Since these are all key nodes in the processing of nociceptive information, it is evident that CB1 receptors are well positioned to regulate pain. An important advance in our understanding the function of the CB1 receptor was the discovery of anandamide,[5] the first endogenous ligand for this receptor and the endocannabinoid counterpart of THC.

CB1 receptors regulate the function of GABAergic neurons, but they are also thought to exert an analgesic effect at the synapse between the first- and second-order neurons in the dorsal region of the spinal cord. How this occurs is unusual because anandamide is synthesized in the terminal of the *postsynaptic* neurons in response to an influx of Ca^{++}. Thus, its synthesis is directly coupled to the activation of the second-order neurons by glutamate after an injury. The newly synthesized anandamide is released into the synaptic cleft where it binds to CB1 receptors on the membrane of the *presynaptic* terminal. This communication from the postsynaptic to the presynaptic terminal is known as *retrograde signaling*. The binding to the receptor elicits a typical conformational change that results in the inhibition of the voltage-gated Ca^{++} channels. We know that this will suppress the release of the glutamate and prevent the activation of the second-order neurons. Anandamide therefore inhibits the transmission of nociceptive information via mechanisms used by GABA, the opioids, and the other transmitters associated with the descending systems we discussed earlier. Because anandamide reduces the efficacy of glutamate in activating the second-order neurons, it will also influence the appearance of LTP and we know that this could have long-term consequences. Anandamide is removed from the synaptic space by a high-affinity transport system and degraded by the enzyme fatty-acid amide hydrolase.

While understanding the function of the CB1-anandamide system in the nociceptive pathway is important, the question is whether it will yield any targets for the development of analgesics. As of yet, there is no evidence that the role of the CB1 receptor in nociception is any different than its function elsewhere in the brain. Consequently, any treatment that increases the efficacy of the receptor will result in the same unwanted side effects due to THC. On a more positive note, subtypes of the CB1 receptor have been reported, and if one of these is associated specifically with nociception it could be an excellent target for development of an analgesic.[6]

Marijuana also relieves pain via the CB2 receptors in the periphery.[7] These receptors are located primarily on cells in the immune system and on cells that mediate an inflammatory response to a lesion. Studies thus far are encouraging because they show that stimulation of CB2 receptors does not produce psychogenic or other adverse side effects. The receptors are activated by anandamide, and there is a reasonably good understanding of how this activation reduces inflammatory pain after an injury. We know that the injury will result in the release of ATP and other agents, some of which will bind to receptors on the terminal of the nociceptive neuron. The binding causes an influx of Ca^{++} that generates action potentials but also activate the enzymes that synthesize anandamide. The anandamide is then released into the interstitial space where it binds to the CB2 receptor on inflammatory cells at the injury site. This is the critical step because the binding results in a signal transduction event that *suppresses* the release of the pro-inflammatory cytokines and other factors from these cells. The anandamide also binds to CB2 receptors on leukocytes, which reduces their migration to the lesion site. Thus, the events triggered by the binding to the CB2 receptor strike at the two main sources of inflammatory pain.

Anandamide makes an additional contribution to analgesia in the periphery by binding to the TRPV1 receptor. We learned previously that activation of the TRPV1 after an injury results in the generation of action potentials that contribute to the transmission of pain information. However, we also know that TRPV1 is rapidly inactivated by the continued presence of ligands, and the presence of anandamide could therefore contribute to this inactivation. It is interesting to note that anandamide is inactivated by COX-2. Thus, anandamide levels should increase in response to COX-2 inhibitors, which would augment their antinociceptive effects.

Even from this relatively brief description, it is obvious that there would be many advantages to developing drugs that either activate the CB2 receptor or promote the synthesis of anandamide or prevent its degradation. At present, preventing its degradation seems to be very promising because inhibitors of the fatty acid amide hydrolase (FAAH) enzyme that breaks down anandamide are reported to have significant analgesic effects on pain due to inflammation.

Cannabidiol (CBD) is another ingredient in marijuana that is receiving a lot of attention because it, too, has significant analgesic and anti-inflammatory activities but without the psychoactive effect of THC. CBD has a remarkably diverse group of effects. Studies indicate that it blocks the activity of the CB1 and CB2 receptors that will reduce the efficacy and potency of THC and anandamide. Nevertheless, it also reduces pain signaling by helping to desensitize the TRPV1 channel and by inhibiting the degradation of anandamide by FAAH. What is of great interest is the evidence that CBD blocks the uptake of noradrenaline, dopamine, serotonin, and GABA into the presynaptic terminal. We know from chapter 8 that maintaining the levels of these neurotransmitters will reduce synaptic transmission at the synapse between the first- and second-order neurons. More research is needed, but if CBD does indeed block the uptake of these neurotransmitters, it could replace the tricyclic antidepressants and serotonin/norepinephrine reuptake inhibitors. A CBD formulation currently available in Canada is marketed under the brand name Sativex, which is currently being used as an adjunctive treatment for the relief of neuropathic pain in adult patients with moderate to severe pain from advanced cancer.[8]

Studies into the analgesic effects of the CB2 receptor and cannabidiol are progressing, but there are several other endocannabinoids that recognize the CB1/CB2 receptors and many more compounds in marijuana to be investigated. Particular attention will be on side effects because the cannabinoid system regulates many processes in the body that have not been discussed here. Another issue is the quality of the various cannabinoids preparations that are becoming available. Nevertheless, given what we know so far, it is obvious that marijuana has great promise as a source of targets for pain management.

Pain can be regulated via the release of intrinsic neurotransmitters at the synapses between the first and second-order nociceptive neurons. But why are so many neurotransmitters necessary? The answer is that each one is attuned to regulate a specific level of pain. Thus, the opioids prevent pain that would result from a severe injury, noradrenalin and serotonin reduce pain in response to mood alterations, and GABA limits most common types of pain by preventing excessive firing at the synapse. We are still not certain of the function of anandamide, the endogenous counterpart of THC, but it could prevent pain to obtain a reward. A good example would be the "high" experienced by runners who stress their bodies to achieve a goal. We will have much more to say about this in chapter 10. One very important implication from all this research is the realization that an analgesic drug that blocks the function of only one of the neurotransmitters will not relieve all types of pain. Even the opiates are ineffective against certain kinds of intractable pain.

Intrinsic Molecules in the Nociceptive Pathway

Other potential targets for drug development are the kinases, channels, and receptors in the nociceptive pathway that contribute to pain. But how does one decide which would be a suitable target? Suitable in this context means a molecule that is not widely distributed among the cells in the body. Nature does not have an unlimited number of options in regulating the many functions necessary for life. Consequently, many of the same enzymes and ion channels are used in cells with different functions. Thus, liver cells contain many of the same kinases as neurons. Even among neurons there is redundancy. For example, suppose a sodium channel in neuron A generates action potentials that result in pain, whereas this same channel in neuron B generates action potentials that travel along a completely different pathway with an outcome unrelated to pain. Obviously this channel would not be a suitable target because any drug that inhibited this channel would interfere with both

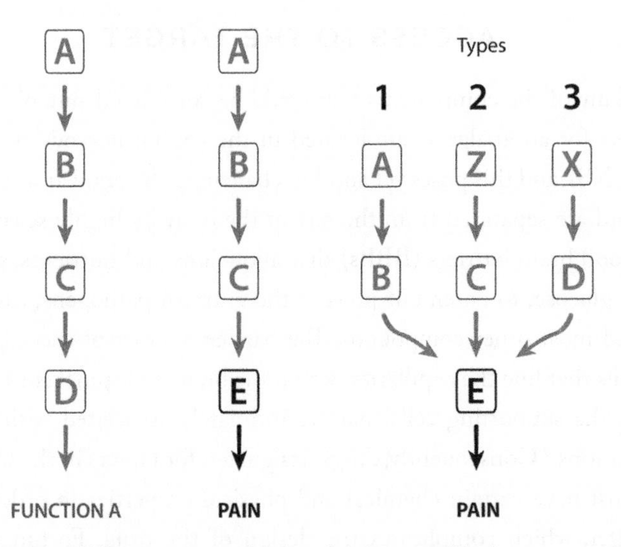

FIGURE 9.1 Selecting a target for drug development. The best target is E because it is unique and closest to the event i.e. pain.

pathways. There are other considerations as well. Figure 9.1 depicts a situation in which a four-step molecular cascade is necessary for an important function in most cells but mediates pain in nociceptive neurons. We can see that a drug that inhibits step A to B will be an effective analgesic, but it will also block that function in all the other cells. The optimal approach, therefore, is to block the step from E because it is unique to the neurons. Another situation shown is when pain arises from different causes. Here inhibiting A, Z, or X will prevent a specific type of pain, whereas inhibiting E will block the pain from all. The obvious conclusion is that scientists must be very careful about choosing the correct target before they begin the arduous process of synthesizing a potential drug.

ACCESS TO THE TARGET

Many of the components that could be considered suitable targets for an analgesic are located in the central nervous system (CNS), and this poses yet another challenge. The brain and spinal cord are separated from the rest of the body by highly selective blood brain barriers (BBBs) that allow ions and nutrients, such as glucose, to enter, but prevent the entry of pathogens, toxins, and most other compounds. The barriers consist of specialized cells that line the capillaries within the brain and spinal cord and by the supporting cells that are intimately associated with the neurons.[9] Consequently, drugs designated for targets in the CNS must have certain chemical and physical properties in order to enter, which complicates the design of the drug. Fortunately, many thousands of compounds have been synthesized and tested over the years and their properties in animal studies have been evaluated. It is now possible to use the data from these studies to guide the design a drug with the expectation that it will breach the BBBs, thereby avoiding extensive and expensive testing.

Once the problem of the BBBs has been resolved, the potential drug can enter the brain, but this poses yet another set of problems. As we know, the brain is a very fine-tuned structure, and any interference with brain functions will have especially serious side effects. This means that the drug must have a very high affinity for the target, which can be difficult to achieve. Consider adverse effects from the opiates, THC, and other inhibitors of the descending systems that we discussed in chapter 8. Other examples are the analgesics designed to block the NMDA receptor, which we know is essential for the development of the late-phase LTP. Several NMDA receptor antagonists (inhibitors) are available, e.g., ketamine, memantine, amantadine, and dextromethorphan.[10] They cross the BBBs and can mitigate pain to some degree, but all have side effects that include hallucinations, dizziness, fatigue, and headache. What all this shows is that even a well-designed analgesic that enters the CNS is likely to have side effects. For this reason, the FDA has recently put more stringent requirements on all agents that act within the CNS; this makes these agents more difficult to design and synthesize and, of course, more expensive.

In contrast to the neurons in the CNS, those in the peripheral nervous system (PNS) have direct access to drugs in the circulatory system. Even with this considerable advantage, however, the development of new analgesics has proven difficult. The TRPV1 receptor/channel, for example, would appear to be an excellent target because it is directly involved in pain signaling. Yet the several attempts by the pharmaceutical industry to develop TRPV1 antagonists as pain therapeutics have failed. Some studies had to be terminated because patients developed significant and persistent hyperthermia, with body temperatures reaching 104°F. A drug was developed that reduced the hypothermia, but it impaired the ability to sense warmth and noxious heat and it was also withdrawn. As can be imagined, these studies were very expensive and failure meant that the investment was lost.

There is still reason to be optimistic, however. As we discussed, the TRPV1 channel is one of five members in a family of channels and the adverse responses were most likely due to the drugs interacting with other members of the family. Consequently, if structural studies can reveal a site in the TRPV1 that is not present in the other family members, then a drug that targeted this site would have few, if any, side effects. Another approach is to determine how anandamide and other ingredients in marijuana desensitize the TRPV1. There are several other components in the PNS that are promising targets for alleviating chronic pain, and these will be discussed later.

DISCOVERY: SELECTION OF CANDIDATES

Once a suitable target has been identified—say a kinase whose activity is considered essential for pain—the next step is to synthesize a potent and selective inhibitor. Potency is a measure of how much inhibitor is necessary to block the activity of the enzyme and the smaller the amount the better. Selectivity refers to the relative ability of the inhibitor to block other enzymes. A highly selective inhibitor will only block the activity of the target and will therefore have fewer side effects. To get an idea of the magnitude of the work that will be required, let's consider E in figure 9.1 to be a kinase and that it has been decided to start a project to synthesize an inhibitor of E as an analgesic. The first step is to synthesize potential inhibitors, and it is not unusual to make a thousand or more compounds, each of which will have to be assayed for efficacy. Assays were laborious and time consuming until the pharmaceutical industry developed robotics that can test thousands of compounds relatively effortlessly.

Only the compounds that block at least 95 percent of E's activity are considered acceptable for further development. The next problem is that most proteins can be grouped into families whose members differ somewhat in function but have very similar structures. For example, E1 could be one member of a kinase family, whereas family members E2 and E3 are present in other cell types that are not involved in pain. Unless the inhibitor shows extraordinary selectivity for E1, the functions in the other cells will be blocked and side effects will result. This usually means modifying each of the acceptable inhibitors to obtain the one that is most selective for E1 versus E2 and E3. Finally, the very few selected inhibitors that met the criteria for potency and selectivity must then be further tested against many of the other five hundred or so known kinases as a final assessment of selectivity. Obviously all of these steps take time and are expensive.

Of the thousands of compounds that entered the process very few will meet the criteria to be considered candidate drugs. But many steps still remain.

PRECLINICAL TRIALS

Each candidate is then subjected to a series of rigorous tests to determine how well it is absorbed into cells and tissues, distributed throughout the body, metabolized, and excreted.[11] Criteria at each step must be met for a candidate to advance to the next level. This culling process is known as Go/NoGo because candidates that don't meet the criteria are eliminated. Subsequent steps assess toxicity, efficacy in pain models, access to the target, and side effects. Selecting the appropriate animal pain model is very important because it must come as close as possible to the human condition. Moreover, there are many factors that need to be carefully controlled because an animal's response to a drug is influenced by its environment and the nature of the tests themselves. Some animals are stubborn and uncooperative, whereas others are very sensitive and overreact.

All of the tests and assessments involve complex protocols that must be approved by an oversight board and are again time consuming and expensive. Very few of the candidate drugs will meet the established criteria and usually only one, the so-called lead compound, is selected to enter the next stage, which consists of the clinical trials.

The identification of the lead compound usually marks the end of the preclinical development. The clinical trials require much more of the lead compound and, as expected, the scaling up of the synthesis is tightly controlled as to the purity of the reagents, the methods employed, and the management of the project. This process is very expensive, so it is at this point that one of the large pharmaceutical companies usually contracts to support the further development of the lead compound.

CLINICAL TRIALS

Once sufficient quantities of the purified lead compound are available, the sponsors file an application with the FDA that contains a complete description of the preclinical findings as well as a detailed plan describing how the drug will be administered, the dosage, what criteria will be accepted as a measure of success, etc.[12] If approved, the drug then enters the human trials, which are designed to evaluate the safety and efficacy of all new drugs. They provide essential oversight to prevent especially serious side effects, as occurred with the drug thalidomide; it was used to quell nausea during pregnancy, but it resulted in thousands of cases of infants with malformed limbs.

The trials typically consist of three phases. In phase one, the drug is administered to up to one hundred volunteers to determine overall safety. If deemed safe, it moves on to phase two, which typically involves several hundred patients who are suffering pain. The goal is to optimize the dosage and assess side effects. In phase three, the drug is administered to hundreds and perhaps even thousands of patients who are in pain to assess effectiveness and identify long-term adverse effects. Overall, the trials can last up to four years and cost hundreds of millions of dollars.

CAVEATS

This brief description of the many steps in drug development gives an appreciation of the time, effort, and expense that goes into getting a drug to market. Since there are so many steps with stringent criteria for passing, it is not surprising that so many of the promising compounds that enter the discovery stage fail to advance and that very few that enter clinical trials actually reach the public. Every failure reflects a loss of many hundreds of millions of dollars, and it is understandable that many pharmaceutical companies are reluctant to engage in the development of new drugs.

The difficulties are further exacerbated when the goal is to create an analgesic. One problem, which we already discussed, is the need to get through the BBBs; this will continue to be an issue as long as the emphasis is on targeting components in the CNS. The second problem is a phenomenon known as the placebo effect, in which pain is relieved when patients believe that the treatment will succeed, even when the treatment has no therapeutic value. This is the basis for the pitch by those who sell pain remedies that have no pain-relieving ingredients. The placebo effect has provided major insights into the nature of pain that will be explained at length in subsequent chapters. Of course, the placebo effect must be taken into account in clinical trials to ensure that the efficacy of the analgesic is actually due to the drug. The trial must include two populations: those who receive the drug and those that receive the placebo. This step greatly increases the cost of the trial, so imagine the disappointment when the results show that the drug performs no better than the placebo.

10

THE NEUROMATRIX

CONSCIOUSNESS, AWARENESS, AND PAIN

The discussions in the early chapters presented a somewhat mechanistic view of pain in which lesion-evoked action potentials propagating within the somatosensory system result in the sensation of pain. According to this view, adaptive changes in the intensity and duration of the pain were due to molecular changes in the intrinsic first- and second-order neurons that comprise the first pathway in this system. We subsequently learned that this perspective had to be revised in light of the evidence that extrinsic pathways, most notably those involving the opioids, noradrenalin, and the endocannabinoids, could regulate this pathway and alter the perception of pain. While these findings expanded the list of targets for the development of analgesics, the pharmacological approach has not been overly successful in treating pain, and some of the pitfalls were chronicled in the previous chapter. Nevertheless, the discovery of the extrinsic pathways has opened entirely new and much broader views of pain. Most important is that pain can be controlled by centers higher in the brain that are responsible for mood, attention, and anxiety. Because we know that these emotional states will vary depending on circumstances, we can now understand why pain is subjective.

The realization that what we feel as pain is influenced by intrinsic circuits in the brain was transformative, and we will now begin to discuss why this has significantly changed our understanding of pain. Of all the possible influences, attention is probably the most important because the primary purpose of pain is to make us aware of a lesion. But awareness is inextricably linked to consciousness, and there really is no consensus as to what consciousness is or how it arises. Although awareness and consciousness might appear to be one and the same, there is a subtle difference. Imagine you are strolling along the street: you are generally conscious of your surroundings—the sky, trees, houses, people, cars, etc.—but when you hear a dog barking, you turn and become *aware* of the dog. Thus, awareness arises when our senses focus our attention on a particular object or event. Put another way, awareness is a state of being conscious of something, rather than everything. Note also that awareness is paired with a sensation—in this case, we are aware of the dog because we hear the barking. Keep this mind because it will become important later in this chapter. While we cannot provide a detailed explanation for how awareness emerges from neural circuits, we can identify the activity of the circuits that culminate in our becoming aware of something. And this is significant because we can willfully ignore the dog. Does this mean we can willfully ignore or be unaware of pain? Only two decades ago the answer would have been no, but based on the recent developments in neuroscience we believe it is now appropriate to say yes.

We have already discussed the phenomenon of stress-induced analgesia in which there is no awareness of pain after a grievous injury. A very intriguing wrinkle in our understanding of the relationship between awareness and pain came from the results of a treatment given to psychiatric patients in the middle of the last century.

Some patients deemed to be irredeemably psychotic or violent were pacified by an operation known as a prefrontal lobotomy. In essence, the front part of both cerebral hemispheres was completely separated from the rest of the brain. In most cases, this operation did indeed reduce violent behavior, but in some it had an additional effect. One patient in particular was working in the kitchen when he touched a hot stove and severely burned his hand. He readily acknowledged that he should be in pain but said he didn't care. Remarkably, the operation somehow disconnected the awareness that he had been burned from the onerous aspect of pain. This idea is certainly difficult to grasp conceptually—that one can be aware of a serious injury without experiencing pain—but it implies that his indifference to pain was due to the fact that the lobotomy separated the somatosensory system from one or more circuits in the brain that assesses the *averseness* of pain. This disconnection was an example of the phenomenon known as *asymbolia*, and it was quite a revelation.[1] Another outcome from the lobotomy studies was that our vocabulary for pain is inadequate. If we can be in pain, but not care, means that we cannot use the term *pain* in the usual sense of the word that connotes at least some degree of suffering. Consequently, we will use "painful," "hurtful," or "experiencing pain" to separate the onerous response to a lesion from the general awareness or perception of the lesion.

In addition to recognizing that painfulness can be dissociated from awareness, we know that pain can be modulated by drugs that alter mood, as well as by the situation or setting in which the pain occurs.[2] To be more precise, we can now postulate that the degree to which we feel pain is not solely determined by the activity of the somatosensory system, i.e., the nociceptive pathway, thalamus, and sensory cortex, but it depends on inputs from neuronal circuits in the brain that qualify the pain depending on circumstances.[3] Thus, hurtfulness ultimately depends on awareness but is shaped by fear, reward, belief, and the memory of past and present events. All of these properties can be grouped into what is known as the *affective component* of pain, as opposed to the *somatosensory component*, which provides the initial perception of the injury.[4] This idea was further refined when, in 1990, Ronald Melzac proposed that these affective properties emerged from centers in the brain that comprise what he called a *neuromatrix* for pain.[5] This proposal had a profound influence on contemporary ideas about pain.

In order to define the role of the affective components in painfulness, we will discuss the function of areas in the brain that are not completely understood. There is no molecular description of awareness or reward, but there is a consensus, based on studies of brain injuries and animal models, that each component of affect can be assigned to a discrete group of neurons in the brain. By analogy, we don't know how vision arises from interactions between the retina, thalamus, and cerebral cortex, but we do know which neurons are involved at each location. Thus, we are justified in ascribing each affective property to a discrete group of neurons that are located in a unique *affective module*. There will be a module for awareness, another for fear, etc. Once we have defined each affective module, we can reasonably link these to the modules that comprise the somatosensory system. What will emerge eventually is an introduction to several new and very exciting avenues for managing pain.

IMAGING THE BRAIN IN ACTION

Acknowledging that there are affective and somatosensory components is one thing, but locating them is another matter entirely. Remember that the characterization of the molecular components in the nociceptive pathway was aided by the relatively simple anatomy of the peripheral nervous system and the ability to use animal models. The brain is infinitely more complex and the areas involved in pain have no obvious counterparts in lower vertebrates. Fortunately, advances in technology have made it possible to watch the brain in action in real time.

Among the several procedures to obtain images of brain activity in a noninvasive way, *functional magnetic resonance imaging* (fMRI) has proven to be particularly important.[6] Based on the same technology as structural MRI that is used elsewhere in the body, fMRI is a refinement that looks either at changes in blood flow or oxygen usage in the brain to detect areas of higher activity. The underlying premise is that neurons activated by a particular stimulus will use more energy than neighboring neurons and will therefore need more oxygen and blood. A limiting factor is resolution, which refers to the size of the area that can be detected. Thus, it is not possible to see individual neurons, merely areas of higher activity. There is also a temporal issue because communication among neural networks occurs in milliseconds whereas acquiring the image takes longer. Although fMRI has limitations, it has been extraordinarily useful in identifying groups of neurons that are consistently found to be involved in particular aspects of affect. Most of the conclusions are drawn from experiments that were carried out in different labs but with similar protocols. We will rely on fMRI imaging, augmented by other procedures, to provide a conceptual understanding of how the neuromatrix functions in the expression of pain.

AWARENESS AND PAIN

fMRI, consistently detected increased activity in several discrete regions of the brain, both in volunteers during the presentation of an acute painful stimulus and in patients who were suffering from persistent pain. One of these regions is comprised of neurons in the *anterior cingulate cortex* (ACC), which is located just beneath the surface of the anterior portion of the cingulate gyrus and adjacent to the corpus callosum on the medial surface of each hemisphere (fig. 10.1A, C). This was highly significant because other studies have shown that neurons in the ACC mediate an awareness of sensations. We can therefore postulate that the ACC is the module within the affective component that makes us aware of pain.

Moreover, the awareness is coupled with the sensation of pain, just as the awareness of the dog barking was coupled with the sensation of hearing. We know that disruption of the cingulate cortex decreases pain because patients with intractable pain who received cingulotomies reported immediate relief from the suffering associated with the pain. They reported that they were aware of the pain, but that it was no longer bothersome, which is reminiscent of the lobotomy results. This suggests that if we could willfully control the activation of the ACC, we could reduce the onerous component of pain. But how are the neurons in the ACC module informed that there has been an injury or other lesion? After all, these neurons are located deep in the brain. The answer comes from imaging that detected pain-associated activity in the postcentral gyrus of the sensory cortex and in the thalamus. These are modules in the somatosensory system, and the increase in activity is exactly what we would predict because we know that activation of circuits in the thalamus mediate the initial intensity of the pain whereas those circuits in the postcentral cortex identify the

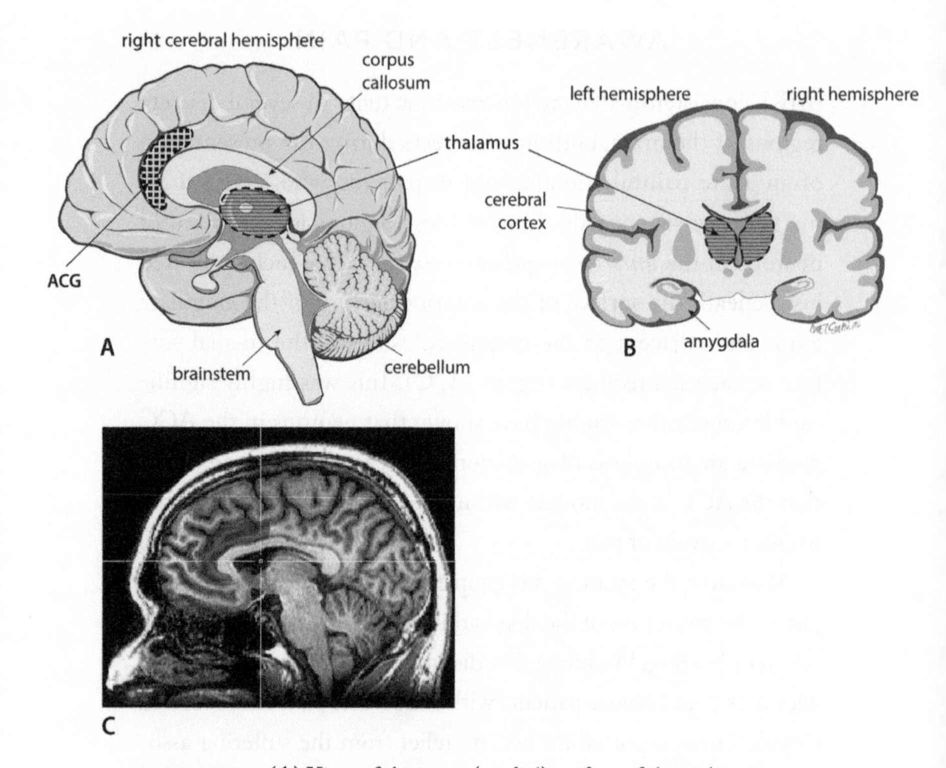

FIGURE 10.1 (A) View of the inner (medial) surface of the right cerebral hemisphere showing the thalamus and the anterior cingulate gyrus (ACG), which is just above the anterior part of the corpus callosum. The anterior cingulate cortex (stippled area) is composed of the neurons that reside beneath the surface of the ACG. (B) Section through the cerebrum showing the right and left thalamus (striped) and the amygdala (black circles) in each hemisphere. (C) fMRI image of the brain in response to a noxious stimulus. The region in the anterior cingulate cortex that is activated is accented.

origin of the pain. Hence, pain is initially perceived via interactions between the thalamus and the sensory cortex. However, we have since learned that perception is linked to awareness; what emerges from the interaction between the thalamus and cortex is actually only a potential perception. The key discovery was made using a procedure that can trace nerve tracts that found direct connections between subsets of third-order neurons in the thalamus and those in the ACC. A direct link between a somatosensory component and an affective component means that action potentials coursing from the thalamus to the ACC can make us aware of an injury.[7]

Showing how we become aware of an injury was obviously important. However, awareness is also hierarchical. For example, our awareness of the dog barking can be immediately superseded by the sound of a siren and an awareness of a fire engine. We pay attention to the fire engine because experience tells us that it is more important than the dog. Similarly, our awareness of the pain will depend on the context in which it occurs and will therefore be influenced by neuronal circuits in the affective modules for fear, reward, etc. Consequently, we need to identify each of these modules and place them in relationship to the ACC.

FEAR AND REWARD

The amygdala is another region of the brain that is repeatedly found to be activated in subjects who are in pain. There is an amygdala in each cerebral hemisphere (fig. 10.1B), and for a relatively small structure its neurons have a major role as a center for emotions. Studies in the 1930s showed that removal of both amygdalae resulted in marked changes in behavior, the most striking of which was a lack of fear.[8] That the loss of such a small group of neurons could so radically alter an essential behavior was a revelation. Also intriguing is that neurons in the amygdala have CB1 receptors, which we know are recognized by THC. This connection explains why marijuana users exhibit diminished fear. Each amygdala has connections to the thalamus on the same side of the brain and thereby receives input from the nociceptive pathway and from all of the senses (except olfaction).

Suppose a child is given a painful injection. This is a traumatic experience and the next time that child sees a needle he or she will exhibit fear and, remarkably, this fear of needles can extend well into adulthood. What happened was that neurons in the child's amygdala retained a memory of the pain from the injection. This is very different from memories of everyday events that are stored elsewhere in the brain. The memories stored in the amygdala do not necessarily have to be of a painful event because the amygdala also serves as a repository of events that were especially threatening or traumatic, such as a fire. Memories such as these can have benefits for survival because fire will then be considered something to be avoided.

On the other hand, the affective component of the matrix also contains a group of neurons that provide positive reinforcement and motivation for behavior based on whether or not there is a sufficient reward for ignoring the hurtfulness. These neurons are housed in the nucleus accumbens.[9] The neurons in this module can be viewed as the yin to the amygdala's yang because the memory of a pain can be overcome if the rewarded is deemed sufficiently worthy. We might have a fear of needles but are willing to overcome the expected pain from an injection of an antibiotic because the outcome—elimination of infectious bacteria—is evaluated as being more important. The reward network is considerably more extensive and has a significant role in the experience of pain. This will be discussed in chapter 12.

The Neuromatrix: Mapping the Components of Pain

By combining the imaging results with those using techniques that define interconnections among the various groups of neurons, we can assemble a map of the neuromatrix (fig. 10.2).[10] The

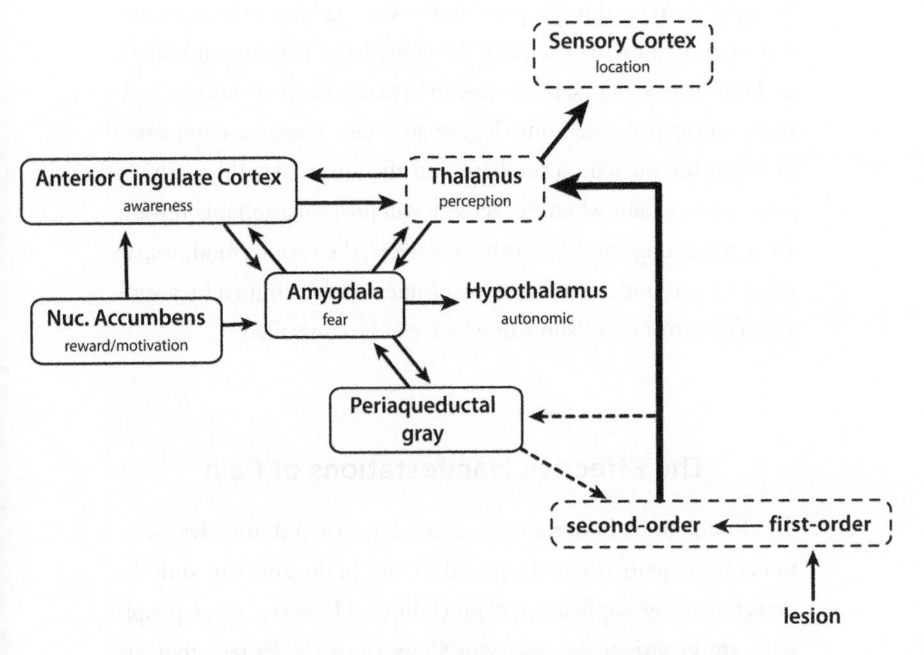

FIGURE 10.2 The neuromatrix. Interactions between modules in the somatosensory system (dashed boxes) and the modules in the affective system (black boxes). The hypothalamus is part of the effective system that connects to the autonomic nervous system.

map identifies each of the modules associated with the somatosensory and affective systems and shows how they are connected by bundles of axons. We already discussed the links between the thalamus and the ACC, and other studies have shown that subsets of neurons in the ACC have reciprocal connections to neurons in the amygdala, which has obvious implications for the awareness of pain. For example, if a person is in a situation that past experience indicates will be painful, signals from the amygdala to circuits in the ACC will indicate fear and the painfulness will be exacerbated, as would be the case with anticipating the pain of a dental drill without anesthesia. Perhaps more important is that pain can also be suppressed via direct inputs from the amygdala to the periaqueductal gray. These will activate the enkephalin neurons and others and interrupt synaptic transmission between the first- and second-order nociceptive neurons. Figure 10.2 also shows a connection between the nucleus accumbens and the amygdala. This pathway can suppress pain when the activity will provide a suitable reward. By determining the relationships between the various modules, we can begin to understand how painfulness is determined by a variety of circumstances, some of which can be controlled.

The Effective Manifestations of Pain

The interactions between the somatosensory and affective components of pain occur deep within the brain and can only be detected using sophisticated procedures. However, most people who are experiencing pain will show signs of distress that are obvious to an observer, and many of these are due to the activation of neurons in the paired hypothalamus (fig. 10.2). Each hypothalamus lies just beneath the thalamus and contains small groups of neurons that communicate with the autonomic nervous system. Remember that this is a motor system that regulates the activity of our viscera as well as glands in the skin, the status of blood vessels, etc. When someone is suffering or under stress, inputs to the amygdala from the ACC are relayed to neurons in the hypothalamus. Their output results in the activation of autonomic nerves that will have an effect, such as sweating, an increased heart rate or tear production, and other physical signs of distress. These are often good indicators of the intensity of pain and clinicians can use these signs to validate a patient's claim to be in pain. More subtle manifestations of stress can be detected using a so-called lie detector when a suspect is being questioned about a crime.

The map of the neuromatrix provides a visual rationale that explains how the affective modules for awareness, fear, and motivation/reward can modulate the painfulness of an injury or inflammation that is communicated via the modules comprising the somatosensory component. Now we will discuss a phenomenon that markedly increases the value of the matrix in understanding pain.

PSYCHOLOGICAL PAIN

The prevailing view is that circuits in the affective component modulate the perception of pain that arises from a lesion. However, what we have learned recently is that these circuits have another purpose, namely to elicit pain that occurs in the absence of an injury or inflammation. The ancient Greeks recognized that suffering could have both physical and psychological causes, and Homer, author of the *Odyssey*, distinguished painfulness from an injury from mental anguish. In modern terms, this distinction is between pathophysiological pain and psychological pain, and there has been strong skepticism in the medical profession as to whether or not the latter exists. Many of those who study or treat pain consider it to be a response only to a lesion. In other words, they deny that activation of the affective component can itself result in pain. Cases in which patients complained of pain where no cause could be found were thought to be hysterical or have some form of mental derangement. Implicit in these diagnoses was a denial that the pain was real and that the patient was truly suffering. Yet there was much anecdotal evidence to refute this position. For example, a patient said that the excruciating pain of passing a kidney stone was nothing compared to the pain he was suffering from the death of his daughter. In another example, a grieving widow spoke of pain that was more intense than any other she had experienced in her life. It was not reasonable to believe that they were not suffering. Moreover, these patients suffered for months, meaning that the pain was chronic. The International Association for the Study of Pain has finally acknowledged that people do experience pain due strictly to "psychological" reasons.

The term psychological pain has now morphed into psychogenic pain, *psychalgia*, or *algopsychalia*, depending on the field of study.[11] It recognizes the existence of brain centers that are responsible for the affective aspect of pain and that activation of these centers can elicit painfulness independent of the somatosensory component. It further acknowledges that the pain can be chronic and that it can be caused by grief, stress, and even psychosocial problems. In addition to serving as a source of pain, psychogenic pain can exacerbate pain that has a pathophysiological source, especially backache.

Given that psychological pain does not have a physical cause implies that the somatosensory pathway is not activated. However, this is not necessarily correct. Severe grief will elicit the release of stress hormones that can result in aches and pains in the body. This type of pain is considered *psychosomatic* and will be discussed in chapter 12. Moreover, suffering could still arise by activation of certain third-order neurons in the thalamus, and we know that this occurs in cases of central pain. If we go back to the map of the neuromatrix, we see that the thalamus and the ACC have reciprocal connections, and it turns out that fMRI scans of women who were grieving the death of a loved one revealed increased activity in these regions of the brain. We have to be cautious about overinterpreting this information, but since the ACC and thalamus are both activated in response to pain from an injury, these findings suggest that physical pain and psychological pain share at least some underlying neurological mechanisms.[12] In the following chapter, we will begin to discuss how this information might be used in cases of chronic pain.

11

THE BRAIN AND PAIN

The first several chapters in this book described the anatomical, cellular, and molecular bases of how pain is perceived in response to an injury or inflammation. In the previous chapter, we brought our knowledge of pain into the twenty-first century by introducing the neuromatrix theory, which incorporates brain circuitry into the pain story. We also learned that chronic pain can have psychological origins, even in the absence of a physical lesion, which means that such pain can no longer be viewed merely as a pathological change in the nociceptive pathway. Finally, we now know that the affective component of the neuromatrix—the neuronal modules in the ACC (anterior cingulate cortex), PAG (periaqueductal gray), nucleus accumbens, and amygdala—integrate awareness, traumatic experiences, and reward, all things that control what eventually emerges as the feeling of pain. Nevertheless, it came as quite a surprise when it was determined that the flow of information between the affective and somatosensory elements of the neuromatrix was not as simple as was originally envisioned.

NONSUICIDAL SELF-INJURY

The communication between neurons in the ACC and thalamus is necessary for the psychological pain resulting from bereavement. But recent imaging studies suggest that the importance of this communication extends beyond grief to include a variety of other conditions that can best be summarized as social distress. The distress can arise from rejection by loved ones, exclusion from a social set, or even an inability to find employment. Most significant is that in some cases this rejection leads to despair so profound that, like bereavement, it results in pain. In an effort to cope, some who suffer such pain turn to nonsuicidal self-injury (NSSI), which is a euphemism for self-mutilation. NSSI is most often associated with adolescent girls, although it might be underreported in males who choose different types of self-destructive behavior. Self-mutilation, usually by cutting or burning, is as horrible as it sounds, but it is not an attempt to commit suicide.[1] Paradoxically, it is an attempt to relieve the suffering from the rejection by dissociating or separating the mind from the feelings that are causing anguish: the physical pain, therefore, is a willful distraction from the emotional pain. Note the terms *willful* and *distraction*—they will be used in another context later. Many people who self-harm report feeling little to no pain, and for some it can evolve into a means of seeking a form of pleasure.[2] Psychologists call this phenomenon "pain-offset-relief" and it is fascinating because it indicates that the somatosensory component of pain can regulate the affective component. The suppression of injury-induced pain is most likely due to the activation of the opioidergic neurons in the PAG, as in the case of stress-induced analgesia.

THE CEREBRAL CORTEX AND PAINFULNESS

The neuromatrix theory was an important advance because it explained how the experience of pain is modulated by neurons responsible for awareness, fear, and reward. Even so, these neurons merely alter the pain. A missing and very essential feature is to understand why pain hurts. Think about the lobotomy patient: he was aware that he had been seriously burned, but he did not care because he was not experiencing pain. Consequently, awareness and hurtfulness involve two distinct neuronal systems. Since this dichotomy is difficult to conceptualize, it forces us to reevaluate what pain actually is. We can begin by agreeing that hurtfulness from an injury is the normal experience unless it is modified by the modules in the neuromatrix. We know that the hurt can be diminished by the reward system or exacerbated by fears emerging from the amygdala. Awareness is also essential, but the lobotomy patient was aware that he was injured, so what suppressed his pain? The accepted explanation is that one or more connections to the ACC were severed by the operation. Thus, hurtfulness is not determined by the components of the neuromatrix but involves inputs from neurons in higher centers of the brain that are involved in cognition. Put simply, these neurons evaluate each sensation and determine which is most important given the many immediate circumstances in which the pain is occurring. The evaluation can be based on the surroundings, expectations, and even beliefs. To begin to understand how all this occurs, we need to learn a little more about the organization of the two cerebral hemispheres.

Each hemisphere is divided into five lobes (fig. 11.1A). Four of these—the frontal, parietal, occipital, and temporal—can be generally distinguished by landmarks on the surface, whereas the fifth,

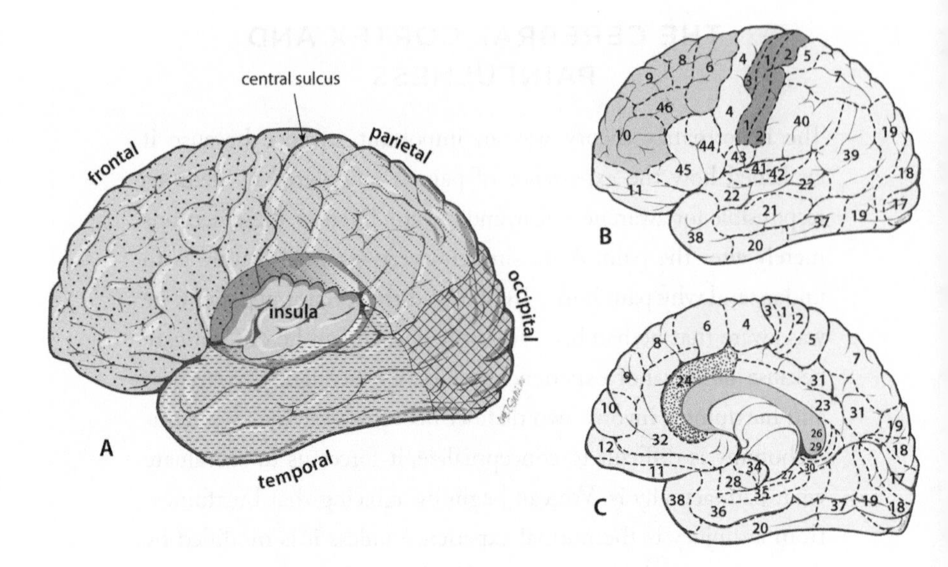

FIGURE 11.1 (A) The five lobes of the left cerebral hemisphere. The insula was exposed by raising the lower border of the frontal and parietal lobes. (B) The outer surface of the left hemisphere showing the map of the cortex as determined by Brodmann. The light gray indicates the regions comprising the prefrontal cortex and the darker gray region is the somatosensory cortex that contains the sensory homunculus. (C) Section showing the inner surface of the right hemisphere and the areas comprising the anterior cingulate cortex (dark gray) and the corpus callosum (light gray).

the insula, cannot be seen because it is tucked under a fold at the lower border of each hemisphere. Immediately beneath the surface of each lobe is the cortex; it consists of the billions of neurons that are responsible for the behaviors that separate humans from lower primates. The neuroanatomist Korbinian Brodmann, in the early 1900s, stained the neurons throughout the cortex and found that he could identify fifty-two regions based on differences in their morphology and pattern of organization (fig. 11.1B). His findings gave credence to the emerging idea that although the brain appears homogeneous, it is actually divided into regions with distinct functions. We have already discussed the area of the cortex that receives sensory information from the body that could be mapped as a homunculus. Advances in the ability to record from brain regions and in imaging have subsequently revealed that the neurons in each region can be further divided into more discrete functional areas and the map now consists of hundreds of subregions.[3] The neurons in each subregion communicate with other cortical neurons, either in the same hemisphere or, via the axons in the large corpus callosum, to the opposite hemisphere. Some are interconnected to neurons in subcortical neurons, such as the thalamus, which sends information to the cortex via the senses. The billions of neurons in the cortex and their trillions of connections process this information and, in ways that are not known, make decisions about how to respond to the surrounding world.

An injury is arguably the most important source of information, and fMRI imaging has identified the modules in the neuromatrix that are engaged when we experience pain. An injury occurs under a given set of circumstances and three groups of cortical neurons evaluate the injury based on our knowledge of these circumstances. The first consists of the neurons located in the ACC (fig. 11.1) that are involved in attention or awareness of a sensation; they were discussed in chapter 10 as part of the neuromatrix. The other two regions (insula cortex and the prefrontal cortex) are especially significant because they each provide a cognitive and evaluative aspect to the experience of pain.

The Insula Cortex

The neurons in the insula cortex (IC) are located deep within each hemisphere (fig. 11.1A) and are subdivided into regions based on their connections to other cortical neurons and the ACC.[4]

The latter connections are especially important. Remember that inputs from the thalamus to the ACC provide sensory information—touch, vision, hearing, taste—that conjures up perceptions as to what is occurring outside the body. These sensations cannot all be given equal priority because survival requires that we focus on what is most important. Thus, the communications between the IC and ACC form a *salience network* in which each sensation's significance is evaluated. It appears that we can only pay attention to one sensation at any given time. Remember how we switched our attention from the barking dog to the siren from the fire engine? Based on previous experience, we considered the siren more significant than the barking dog. Significance is also shaped by connections from other cortical regions that impose a subjective evaluation on a particular sensation depending upon mood. For example, the sensation might elicit a feeling of disgust, fear, or even happiness. The neurons of the IC seem to be particularly attuned to information about an injury or other lesion because neuroimaging consistently shows that IC neurons are activated by noxious stimuli, and electrical stimulation of the IC evokes painful sensations, such as a pinprick or burning.

We have now added a new layer to our understanding of pain. The links between neurons in the thalamus and those in the ACC make us aware of a given sensation, but it is the interaction between the ACC and IC that determines which particular sensation warrants attention. Moreover, the attention arises because these interactions result in a degree of hurt. Of course, we would expect that information from the thalamus about an injury would have priority and result in enhanced attention, but this is not quite correct because we know from experience that under certain circumstances another stimulus can distract from pain. This could be a caress, music, a foul odor, or anything that draws our attention. In contrast to its role in distraction, the IC is also activated when there is an anticipation of pain. Thus, the IC has a central role in determining whether or not pain will hurt, which has obvious implications for pain management.[5]

The Prefrontal Cortex

As its name implies, the prefrontal cortex (PFC) comprises the cortical neurons found in the frontal region of the frontal lobe (fig. 11.1A). Its functions are probably the most important for what separates us from other primates, yet they're among the least understood. The neurons in the PFC are highly interconnected with much of the brain, including extensive interactions with other cortical, subcortical, and brain stem sites. As such, the PFC is an essential part of a vast network that differentiates among conflicting thoughts and determines, by predicting potential outcomes, which one would be expected to achieve a given goal. We will soon learn that expectation is linked to reward and motivation, both of which are very important in regulating painfulness. Making a decision also relies on a memory of previous events, so the neurons in a subregion of the PFC (area 46, fig. 11.1B) are especially important because they evaluate the potential significance of pain by comparing the knowledge (cognition) of present circumstances to memories of past events.[6] This reasoned response to an injury or other type of lesion is quite different than the memories stored in the amygdala, which provides a reflexive response to situations that had been traumatic in the past.

In summary, then, neurons in the IC and PFC provide a third layer to the experience of pain. The first, the somatosensory system, encodes basic information about the location of a lesion and the potential intensity and duration of the pain. The second, the affective component of the neuromatrix, draws attention to the lesion and modulates pain based on certain previous experiences. The third involves a subjective evaluation of the injury that imparts relevance based on knowledge, context, and extenuating circumstances. We can therefore propose that the hurtfulness of pain arises from the cumulative action of neurons in the ACC, IC, and PFC. The IC and PFC's contributions are especially important because they show that the experience of pain depends on higher brain functions that might be willfully controlled.

MASOCHISM AND CONTEXT

The term mashochism broadly encompasses any behavior in which the experience of pain is diminished by the activity of circuits in the brain. Consider the example of an athlete willing to accept pain in pursuit of an important prize, i.e., "no pain, no gain," which involves the affective modules for awareness, reward, and motivation. Far more complex are situations in which pain is modulated by context. Sexual masochism is a condition in which a subject requires some form of painfulness (or abject submission) to achieve pleasure from sex.[7] When a group of masochists were voluntarily exposed to a stimulus that was rated as painful, the intensity of the pain did not differ from that of a control group, and fMRI images from the brain showed that the same regions that we have discussed were activated. However, when the same stimulus was presented to the masochists as they were viewing arousing masochistic images, the intensity of the pain was significantly reduced compared to the control group. Scans also showed that there was increased activation in the ACC and anterior IC compared with controls. We know that the neurons in these regions communicate with one another to provide salience, meaning that the images of masochistic activity were evaluated and determined to be important for achieving arousal and diminishing the pain. Thus, the pictures changed the context in which the stimulus was given. Notable was that the masochist group showed no increase in the activity of areas in the brain involved in the processing of reward.[8] In addition, the reduction in pain was not associated with any activity in the PFC. This lack of pain reduction was interesting because it contrasts with a study of how a group of deeply religious subjects responded to a painful stimulus. That study showed a significant diminution of pain when the subjects viewed pictures of particular religious significance relative to pictures that were deemed to be of little significance. The modulation of painfulness was associated with increased activity in the PFC. As with the masochists, the pictures changed the context in which the painful stimulus was applied; in that way, the evaluation of the context involved circuits in the PFC, not the IC, that were thought to be linked to powerful memories of positive religious experiences. The studies of the masochists and religious subjects show how context can engage different regions of the brain to attenuate pain. We are now going to discuss several situations in which context can be manipulated to reduce pain.

THE PLACEBO EFFECT

Stress-induced analgesia is certainly a very dramatic example of how the brain can regulate pain, but in its most extreme form it is basically a reflexive response to a life-threatening situation. A much more profound and clinically valuable example of how the mind controls pain is the placebo effect, which is a fascinating phenomenon that occurs when pain is relieved by a sham treatment.[9] A placebo can be as diverse as a fake pill, a saline injection, or even a ritual. Anecdotal accounts in popular literature and studies of patients throughout history clearly indicate that pain can be ameliorated by treatments that have no direct therapeutic effect. We have all heard stories about charlatans who have profited from peddling "magical" elixirs that don't actually contain any pain-relieving ingredients. Similarly, shamans and their like acquired power by convincing people that pain could be relieved by secret rituals known only to them. Naturally, there were many skeptics in the medical profession who argued that if the pain could be relieved by a sham procedure, then the person claiming to be in pain was faking. The issue was put to rest when careful studies clearly showed that approximately 33 percent of patients in pain do obtain relief from a sugar pill.[10] Once a placebo was accepted as a valid way to relieve pain, it became important to determine how pain is suppressed by a treatment that does not have any connection to the nociceptive pathway.

The Setting

It turns out that whether or not a placebo is successful in attenuating pain depends on many factors, including who is giving the placebo, such as a doctor or a stranger, knowledge of the treatment, verbal encouragement, and mood. In general, a placebo is much more likely to relieve pain if the patient believes that the treatment will be successful. Thus, if a patient has been taking a pill that eliminates his pain, the pain will continue to be relieved if the patient is unknowingly given a pill that looks the same but is a placebo. In contrast, if the patient is skeptical of the success of the treatment, the placebo is much less likely to be successful. Success then is linked to the patient's *knowing* that the pill worked in the past and an *expectation* that the pain will be relieved. We already know that these two properties arise from circuits in the IC and PFC. The logical next step is to see what parts of the brain are activated in patients given a placebo.

The Placebo Effect and Brain Activity

fMRI imaging of patients exhibiting a successful placebo effect has provided a reasonably good snapshot of which regions of the brain are active in suppressing the pain.[11] The images consistently detected increased activity in the PFC, the nucleus accumbens, and the periaqueductal gray and reduced activity in the thalamus, ACC, somatosensory cortex, amygdala, anterior IC, and the spinal cord. Other studies showed that the PFC is connected to the ACC and the PAG, which is linked to the nucleus accumbens. Given what we have learned about the role of these regions in pain, we can piece all this information together in a narrative that explains how a placebo can be effective. The subject receiving the placebo is in a setting that promotes the belief that the treatment will be a success. The ascendance of the belief involves the activation of cortical neurons in the PFC, which send signals that radiate to other brain centers, including the ACC and nucleus accumbens. Remember that the PFC and ACC were also involved in diminishing pain in religious subjects seeing images of particular relevance.

The nucleus accumbens is part of the reward system and will motivate the subject to take the pill. The input to the ACC reduces the activity of its neurons, which we know will reduce the initial awareness of the pain. Inputs from the PFC to the PAG result in the activation of the opioidergic neurons whose axons descend to the spinal cord, where the release of the opioids prevents synaptic transmission at the synapse between the first- and second-order neurons in the nociceptive pathway. This prevents lesion-induced action potentials from ascending to the brain and accounts for the reduced activity in the thalamus, somatosensory cortex, and ACC. Admittedly, some of this is conjecture, but the essential role of the PAG is corroborated by the finding that the placebo effect is blocked by naloxone, which, remember, blocks the opioid receptor.

These studies reinforce the earlier finding that components of the neuromatrix do not comprise a totally enclosed system but one that can be governed by decisions imposed from higher centers. The studies of the placebo effect are particularly important because they show that the activation of the PAG is controlled by circuits in the PFC and IC. Since these circuits exert volitional control over the PAG, it should be possible to suppress pain by willfully activating the PFC and IC.

In addition to its effects on pain, a placebo can activate brain centers that connect to the hypothalamus, which we discussed earlier. The output from the hypothalamus will drive autonomic functions, and participants in placebo groups exhibit changes in heart rate and blood pressure. The connection between the placebo effect and bodily functions has led to the idea that a placebo can eliminate the cause of the pain. Unfortunately, many studies have determined that this mind-body interaction does not occur.

HYPNOSIS

One takeaway from the discussion on placebos is that the experience of pain can be controlled by regulating the activity of neurons in the ACC and PAG. Reducing activity in the ACC diminishes pain by lowering awareness, whereas activation of neurons in the PAG results in the release of the endogenous opiates in the spinal cord that shuts down the nociceptive pathway to the thalamus.

Many ancient cultures, especially those in the Orient and India, recognized that some people could be put into a trance-like state in which they had a diminished awareness of reality, so they used various forms of meditation to reduce their stress and improve their health. Franz Mesmer, a German physician, popularized a version of these practices in the mid-1800s in Europe in which he was able to able to induce a state in which people were "mesmerized." Mesmer is now considered to be the father of what is now called hypnosis.[12] Mesmer used a variety of approaches to hypnotize his subjects and often employed music, which he thought was a way of bypassing the conscious mind. More likely, the music was a distraction that made the subjects acutely aware of something rather than everything, which is why dentists often play music in their office. Hypnotism was brought into medical practice but soon fell out of favor because it became associated with parlor games and magic acts. Hypnotism has seen a recent revival (as hypnotherapy) to treat pain, anxiety, insomnia, and other problems.

Unfortunately, only about 10 percent of the populace can enter a deep hypnotic trance. Most can achieve intermediate levels, but 10 percent cannot be hypnotized at all. Those capable of deep hypnosis enter into a state in which there is an increased focus of attention and a reduced awareness of surroundings.

A group at Stanford headed by Dr. David Spiegel studied patients under hypnosis to determine which areas of the brain are involved.[13] Awareness is linked to the ACC, and indeed they found that brain images from subjects under deep hypnosis, like those believing in a placebo, showed reduced activity in the ACC. What was most intriguing happened when the hypnotized subjects were challenged with a pain paradigm. When told that they were going to experience pain, there was an increase in ACC activity, whereas when told that the pain would not actually hurt, there was reduced ACC activity. These changes also correlated with the activity of regions in the prefrontal and other cortices, which could explain why hypnotized subjects exhibited enhanced ability to focus on a particular object or idea. These studies were especially important because they showed that an increase or decrease in experiencing pain correlated directly with the activity of neurons in the ACC.

ACUPUNCTURE

Acupuncture originated in China at least two thousand years ago as a treatment for disease. It consists of inserting very thin needles to various depths at various points along twelve meridians that are believed to course through the body.[14] Each meridian is thought to convey a life force, qi, which has two components: the yin (passive and dark) and the yang (active and light). Each meridian is associated with a designated set of organs. Disease or pain is caused when there is an imbalance between the two forces; the goal of acupuncture is to manipulate the meridians to restore balance. Acupuncture initially spread to Japan and India and is now practiced throughout the world. There is no anatomical structure or landmark that defines a meridian. Hence, the placement of the needles will vary among practitioners, as will the depth, and in some practices the needles are electrified. Despite the obvious subjectivity that is introduced by all this variability, evidence shows that acupuncture is more effective than a placebo, especially for lower back pain. As might be expected, imaging studies show that many areas of the brain are involved so the effect cannot yet be assigned to any modules in the pain matrix. Nevertheless, there is some evidence that acupuncture relieves pain by causing the release of opioids. Thus, acupuncture, like a placebo, works via the activation of neurons in the periaqueductal gray.[15]

MEDITATION

Hypnosis and the placebo effect show that the brain can be deluded into ignoring pain. However, both these processes are elicited by outside agents: the hypnotist or the person administering the placebo. Meditation, on the other hand, is a method whereby practitioners are their own agents, which makes meditation much more widely available as a way to relieve pain. Meditation has been practiced for thousands of years by Buddhist monks with the claim that practitioners can dissociate painfulness from an awareness of an injury. We should recall that this is exactly what the lobotomy patients reported. In the next chapter, we will explain how our knowledge of the neuromatrix has provided support for the ability of various meditative practices to radically alter the experience of pain and how these practices are changing how pain can be managed without drugs.

12

THE MIND REGULATING THE MIND

THE PAIN MATRIX

Alleviating pain has been a goal throughout recorded history, and Western cultures in particular have largely depended on pharmacological agents—elixirs and opiates in ancient times and, more recently, drugs specifically designed to attack molecular targets in the nociceptive pathway. While there is still hope that an effective analgesic for chronic pain will emerge, for many people the pain persists and seems untreatable. Fortunately, given recent advances in neuroscience, there is now sufficient reason to consider nonpharmacological approaches to controlling pain. We can be optimistic because of what we have learned about the three essential elements that regulate pain: the somatosensory and affective modules of the neuromatrix and the cognitive centers in specific areas of the cerebral cortex. Since these modules and systems don't act in isolation, but as components of a vast interconnected network, we can now group them together into a pain matrix (fig. 12.1). This extension of the neuromatrix reflects that the modules in the affective system are regulated by the activity of circuits in the prefrontal and insular cortices that we discussed in chapter 11. In this chapter, we will explore how these

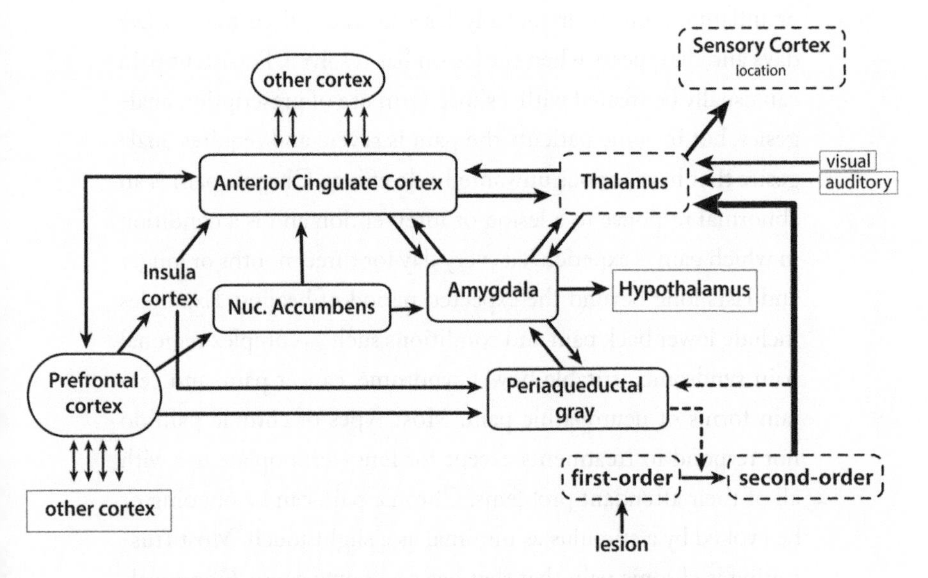

FIGURE 12.1 The pain matrix. Consists of interconnections between regions of the cortex and modules in the affective system (black boxes) and somatosensory system (dashed boxes) that comprise the neuromatrix. Neurons in the hypothalamus activate the autonomic nervous system that controls the body's responses to pain, such as grimacing, tearing, or sweating.

findings can be exploited to begin a new era in pain management in which modules in the matrix are willfully manipulated to manage persistent and chronic pain.

SEVERE LONG-LASTING PAIN AND CHANGES IN THE PAIN MATRIX

We need to be more precise in our terminology and will therefore define two types of long-lasting pain. Persistent pain is a normal response to an intervention, such as surgery, or to a serious injury or inflammation. It predictably lasts no more than four to five days and disappears when the lesion has resolved. Persistent pain can usually be treated with a short-term use of prescription analgesics, but in some patients the pain is severe and requires analgesics that have serious unwanted side effects. Chronic pain is an abnormal response to a lesion or intervention and is a condition in which pain is experienced every day for three months or longer and lasts long beyond the expected period of healing. Examples include lower back pain and conditions such as complex regional pain syndrome, irritable bowel syndrome, cancer pain, and certain forms of neuropathic pain. Most types of chronic pain do not respond to treatments except for long-term opiate use with all of their attendant problems. Chronic pain can be ongoing or be evoked by a stimulus as minimal as a slight touch. Most frustrating is chronic pain that that has no known cause. Fibromyalgia, for example, is a musculoskeletal disorder characterized by chronic widespread pain and tenderness (hyperalgesia and allodynia) without evidence of peripheral tissue abnormalities.

The good news is that imaging studies of patients suffering from either chronic lower back pain or fibromyalgia indicate that the pain is due to abnormal activity in one or more of the neuronal circuits in the pain matrix.[1] This means that we do not have to look elsewhere in the brain for the cause. Most common are increases in the activity of the insula cortex (IC), prefrontal cortex (PFC), anterior cingulate cortex (ACC) and amygdala (fig. 12.1). Activation of the PFC in particular seems to correlate with the intensity of chronic back pain, and we know that an increase in activity in the amygdala will add an element of fear. There are also some indications of changes in the interconnections between the components of the matrix in chronic pain patients. The very good news, however, is that it appears that these abnormal changes in activity and connectivity are reversed when chronic pain is successfully treated.

Although these studies seemed to exclude a role for the nociceptive pathway in chronic pain, they were carried out on patients in whom the pain was already well established. In the following chapter, we will discuss how blocking certain sites in this pathway can prevent pain from transitioning to the point that it becomes chronic. Now, we want to focus on how interactions between the various modules in the pain matrix influence the experience of pain.

THE MODULATION OF PAIN BY THE AFFECTIVE AND COGNITIVE COMPONENTS OF THE PAIN MATRIX

We have learned that the nociceptive pathway transmits information regarding the severity of a lesion to the thalamus and then to the homunculus in the sensory cortex. This information is translated into signals that encode the intensity and duration of the pain and the location of the lesion. In addition, subsets of neurons in the thalamus relay these signals to the affective centers in the brain that in turn communicate with the cortical systems responsible for higher behaviors (fig. 12.1). This description of pain would have been unthinkable a few decades ago. What it means is that the interplay between all of these systems will ultimately determine the awareness of a lesion and whether, and to what degree, we experience pain.

PSYCHOSOMATIC PAIN

An important question is whether the interplay is primarily in one direction—that is, from the somatosensory pathways to the affective modules—or can it also occur in the opposite direction? This interesting possibility has resulted in the belief among some practitioners of pain management that the brain (psyche) can directly cause pain in the body (soma) and that it can occur in the absence of any external pathology. To reframe the issue in more contemporary terms, they claim that chronic pain can arise exclusively from the activation modules in the pain matrix. Whether or not this happens has obvious implications for the treatment of pain and has led to considerable conflict among those in the various medical professions who deal with patients in pain. We know that the brain communicates with the body by stimulating the hypothalamus, which then activates the autonomic nervous system (ANS). As discussed in chapter 7, the ANS regulates basic bodily functions, such the heartbeat and the rate of intestinal peristalsis, and controls the release of stress factors that influence metabolism and the function of the immune system. Consequently, anxiety, extreme anger, or stress that elicits excessive activation of the hypothalamus can cause pain in several ways: by increasing the release of acid in the stomach, potentially causing ulceration, by constricting blood vessels, thereby depriving nerves of blood (ischemia), or by stimulating the immune system resulting in an inflammation. The most common types of psychosomatic pain are ischemic headaches and ulcerative colitis. Overstimulation of the hypothalamus can also exacerbate pain from already existing conditions, as in rheumatoid arthritis. Most important is that these painful psychosomatic disorders will manifest as physical disruption or damage to tissues in the body and therefore conform to what we know about the function of the nervous system. The aches experienced by those who are grieving can be considered a type of psychosomatic pain. In recognition of these findings, the American Board of Medical Specialties and the American

Psychiatry and Neurology Board approved a specialization in psychosomatic medicine in 2003.

The more vexing question is whether the brain can experience pain coming from a specific structure in the body when there is no pathology? Patients who suffer from this type of pain are often stigmatized as hysterics or hypochondriacs. The premise underlying this type of psychosomatic pain derives from pioneering studies by Sigmund Freud (1868–1939), the father of psychoanalysis, who recognized the complexities of the brain and formulated theories to explain human behavior.[2] He postulated that the brain functions at various levels of awareness. Most important when considering pain are conflicts between the conscious and unconscious mind. Freud viewed the unconscious mind as a repository of primitive urges and memories of traumatic events, which must be repressed because their expression as rage or other destructive behaviors could result in societal disruption and potential ostracism. A function of the conscious brain is to keep these emotions in check, and it does this by eliciting painful symptoms in the body as a distraction. Proponents of this theory believe that it explains many types of chronic pain and that these pains can be relieved by psychoanalytical-based therapies that relieve the repression.[3] These ideas have not been widely accepted by the medical community. On a purely conceptual level, it is difficult to believe that substituting a debilitating pain for an expression of rage is a wise strategy for survival. In addition, an important factor in this type of psychosomatic pain is that it is localized to a specific area of the body. Recognizing the source requires direct activation by the brain of those areas in the sensory homunculus, yet pathways that could mediate this activation have yet to be found. Finally, and most importantly, we cannot ascribe functions to the subconscious because we simply do not know what it is or where it resides. There are no fMRI images of the subconscious mind in action, so it is purely a theoretical construct. The value of defining and understanding the functions of the modules in the matrix is the hope that we can somehow alter the activities of the key modules to reduce pain. Some adherents of Freud's theory have had success in alleviating pain, but as with most alternative approaches, it must be shown that these were not due to carefully selecting patients so as to elicit a placebo effect.

In summary, there is no question that somatic pain can be caused by the excessive activity of brain circuits linked to the hypothalamus and that this type of psychosomatic pain can be treated with drugs or counseling to reduce the anxiety or other precipitating causes. We will revisit later instances of so-called central pain in which the pain arises exclusively from the activation of the thalamus. This is very rare, and there is no evidence at present to support the idea that most types of chronic pain are due to the repression of primitive emotions in the unconscious mind.

Our goal now is to learn how the power of the mind can be harnessed to mitigate pain; this path entails identifying the modules within the matrix that are essential for the expression of persistent and chronic pain.

PAINFULNESS IS DIMINISHED BY KNOWLEDGE, BELIEF, AND REWARD

Among the most important recipients of information from the thalamus are the circuits in the ACC (fig. 12.1). Imaging studies showed increased neuronal activity in the ACC after an injury and decreased activity in patients who were successfully treated for pain using a placebo or were under hypnosis (table 12.1). Although it is clear that the ACC is important for the awareness of an injury, we know that awareness does not equate with

TABLE 12.1 CHANGES IN ACTIVITY IN THE COMPONENTS OF THE PAIN MATRIX UNDER THE CONDITIONS SPECIFIED

	Injury	Placebo	Hypnosis	Anticipation	Attention
PCG	increase	decrease		increase	
Thalamus	increase	decrease			decrease
ACC	increase	decrease	decrease	increase	increase
IC	increase	decrease	increase	increase	decrease
PFC	increase	increase	decrease		increase
PAG	increase*	increase			increase
Amygdala	increase**	decrease			
N. accumbens		increase			

Open boxes indicate that the response component was not monitored. PCG: postcentral gyrus; ACC: anterior cingulate cortex; IC: insula cortex; PFC: Pre-frontal cortex; PAG: periaqueductal grey. *After extreme injury or stress. **When fear is involved.

suffering. Rather, it is the inputs to the ACC that render a lesion painful, so we need to determine the origin of these inputs in order to learn how to mitigate the pain.

In the previous chapter, we presented evidence obtained from hypnotized subjects that activity in the ACC is important for the hurtful aspect of pain. Now let's consider what happens when hypnotized subjects were informed that they would be given a painful stimulus but that it would not hurt. Remarkably, they reported experiencing little pain and concurrent imaging showed a decrease in ACC activity. Thus, it appears that *knowledge* prior to delivering the painful stimulus somehow altered the activity of the ACC and the painfulness. Knowledge, of course, is a property of neurons in the cerebral cortex and it was telling, therefore, that there was an increase in the activity of neurons in the IC in the hypnotized subjects who reported diminished pain (table 12.1). Remember that an interaction between the ACC and IC is important in determining whether pain will be experienced. ACC activity in response to a painful stimulus was also reduced when patients were treated successfully with a placebo (table 12.1). In this case, the patients at some level *believed* or *expected* that the placebo would work and this was associated with an increase in the activity of circuits in the PFC and reduced responses to a noxious stimulus in the ACC and other areas of the matrix that correlated with the relief of painfulness.

The activation of inputs to the ACC from the IC and PFC during placebo and hypnosis-induced analgesia points to the importance of the cerebral cortex in modulating pain. One caveat is that the roles of the IC and PFC are undoubtedly far more complex than depicted here. We know that these areas communicate with other cortical circuits, and each other (table 12.1), so treating each one as an isolated system is an oversimplification. They are components of the vastly more extensive cognitive network that makes decisions based on inputs from sensory systems, memory banks, and emotional centers. The PFC in particular is involved in intelligently guiding our thoughts and emotions to adapt to current circumstances. We don't know all the details, but we can still postulate that activation of circuits within either the IC or PFC will have a significant impact on the experience of pain.

Table 12.1 shows that the neurons in the nucleus accumbens were activated in patients whose pain was successfully reduced with a placebo. As discussed previously, these neurons have a major role in assessing the value of a proposed action and they have extensive connections to circuits throughout the brain, including the cognitive centers in the PFC and the awareness networks in the ACC. A primary function of all this connectedness is to determine whether or not achieving a specific goal will be sufficiently rewarding to justify the effort that will be necessary. Scientists who study behavior have found that humans have an innate value system that values an achievement more if it requires an effort. Thus, we gain more satisfaction from a high score in math if we studied a great deal than if the score was attained easily. Pain can also be a source of reward and motivation and the nucleus accumbens contributes to the decision that bearing some pain will be acceptable because the reward is believed to be sufficiently important. If we examine the previous sentence carefully, we see that there are actually two components to this decision—acceptance and belief. Relating this to our own lives, we might be willing to accept pain from, say, lifting weights during training, or sprinting to the finish line if we believe the race has great value. Sometimes the decision has to be made very quickly. If we pick up a very hot cup, we will drop it to avoid being burned. However, if the cup is part of a very valuable set, then we will bear the pain and gently put the cup on its saucer. In this case the reward might be to avoid embarrassment or to be proud that we saved the cup.

SUFFERING AND BELIEF

Nowhere is the connection between acceptance and belief more important than in the role of suffering in religion.[4] Religious beliefs are among the most powerful influences on behavior, and ritual self-injury has a long history in many religions as a rewarding way to atone for one's transgressions. Adherents are willing to accept pain in the belief that it will bring them closer to their God. But how much suffering are they willing to tolerate? This question is relevant to a discussion of analgesia because all religions have stories of martyrs who are willing to die or be tortured for their faith. The motivation (reward) to accept death rather than capitulate must be very powerful; scholarly interpretations of ancient texts suggest that although the martyrs felt pain, they considered it redeeming and did not really suffer. In other words, their belief in their deity was so strong that they accepted the pain because they believed that they would be rewarded in heaven. This post hoc view of martyrdom was promulgated by theologians, such as St. Thomas Aquinas, and was depicted in many paintings of Christian martyrs that evince no expressions of suffering. Of course, this view is somewhat tainted because the theologians claimed that the martyrs didn't suffer because they were infused with divine grace. Nevertheless, martyrdom without suffering appears in other religious traditions as well and it is considered credible enough for some scholars to argue that the martyrs were indifferent to pain because their belief was so powerful that their acceptance of the physical trauma resulted in the release of endogenous opiates, which of course is stress-induced analgesia.[5] Whether or not these interpretations of martyrs' fates are correct is an open question, but if true, then it would indicate that cortical centers involved in decision-making can reduce or eliminate pain by acting on downstream components of the reward and belief networks. The PFC is part of this reward network, and this supports the idea that chronic pain patients might be able to train neurons in the PFC to accept the pain as a reward for some cause. What is especially interesting are links from the PFC to the PAG and the nucleus accumbens. We will have more to discuss about this and the acceptance of pain later on.

Whereas there is much we don't know about the reward system, we do know it is a powerful motivator and that the nucleus accumbens makes an essential contribution to this process. Another source of motivation is the pursuit of pleasure, which is a universal human attribute, and neurons in this nucleus reinforce or incentivize behaviors that lead to pleasure. For this reason, the nucleus accumbens is widely considered to be the center that promotes hedonism. The seeking of experiences that are deemed pleasurable or gratifying can be so powerful that it sometimes overrides rational thought, which can be disastrous. A good example is that the pleasure derived from smoking opium, injecting heroin, or snorting cocaine can overcome concerns from the cognitive cortex about potential future consequences. Consequently, the nucleus accumbens has a significant role in promoting behavior that leads to addiction.

PAIN IS EXACERBATED BY UNCERTAINTY, FEAR, AND STRESS

Now we need to shift focus from pleasure to discuss the unsettling evidence that activation of some elements in the pain matrix actually enhances suffering. Just as reduced ACC activity correlates with a diminished experience of pain, studies have shown that anticipating pain increases ACC activity and exacerbates painfulness. We noted earlier that hypnotized patients reported a reduction in painfulness when told that a stimulus would not be painful. A very different outcome resulted when the same hypnotized subjects were told that the stimulus would hurt: they experienced heightened pain that was accompanied by increased activity in the ACC. In another study, subjects given uncertain expectations about a painful stimulus developed transient increased responses to a nonpainful stimulus and increases in activity in the ACC and IC (table 12.1). This was especially true if there is a traumatic memory of the anticipated event in the amygdala, which is connected to the ACC. Consider what happens to many patients when they hear the whir of the dentist's drill: the sound engenders fear and stress, both of which actually lead to an increase in painfulness when the drilling begins. Another important factor in the experience of pain is our emotional state, with negative emotions enhancing pain-evoked activity in the ACC and IC even in the absence of an actual physical stimulus.

Stress might be unwanted, but it is a normal response to a situation that might be threatening and therefore requires attention. Circuits in the brain are activated in order to consider how to deal with the stress and, as we all know, this often leads to confusion when the stress is overwhelming. Thus, the brain does not necessarily deal well with stress, especially when the cause persists, and this often adds an element of anxiety. Clinicians see much more stress-induced anxiety during crises where individuals have little control over the outcome. Senses are heightened and relaxation becomes difficult. Being in pain for long periods is obviously very stressful, and there are many reasons to believe that the suffering experienced by patients with persistent or chronic pain is enhanced due to a fear of the pain and the stress that it places on the quality of their lives. In fact, the psychological term for it, *pain catastrophizing*, encapsulates many of the negative emotions that contribute to the experience of pain. A goal of psychologists and psychiatrists who treat chronic pain patients is to create a more positive outlook.[6]

Merely saying that stress is an important factor for the experiencing of pain does not provide an explanation as to how this occurs. The hypothalamus is one route, and we have learned that excessive activation of this module contributes to a variety of psychosomatic pains. A more specific cause comes from studies showing that persistent stress causes an elevation in the levels of cytokines, especially interleukin-6, which we know from a previous chapter are responsible for the pain from an inflammation via the somatosensory system.[7] Evidently, the increase in cytokine levels also leads to the activation of affective components in the pain matrix. In one situation, the increase was associated with bereavement and an increase in the activity in the ACC and the PFC. Drugs that inhibit the stress-induced increase in cytokine levels could therefore be useful adjuncts in controlling the emotional component of pain.

What we have just learned is the almost bewildering number of mental processes that determine the degree of painfulness that we experience. Reward, acceptance, knowledge, and belief can diminish the experience of pain, whereas stress, fear, anxiety, and emotional state can worsen the pain. What is quite remarkable is that all of these influences are mediated by networks within the IC, PFC, and ACC. The key to controlling pain is to control these networks.

THE SELF-REGULATION OF PAIN

Painfulness can be relieved via hypnosis or a placebo, and the analgesic effect derives from the concerted activation of components of the pain matrix that mediate reward, belief, etc. Unfortunately, very few people can reach the deep hypnotic state necessary for the analgesia and the success of a placebo depends on a complex relationship between the patient and the physician. Far better would be to have patients relieve their pain by willfully exerting control over the activity of modules in the pain matrix. The goal would be to learn how to activate pathways that relieve suffering and to inactivate pathways that lead to fear and anxiety. A most promising strategy to attaining both goals is to take advantage of what we have learned about *diversion*. If dwelling on pain exacerbates the painfulness, then diverting attention away from the pain might remove the hurtfulness. Let's then discuss what we know about how we pay attention to anything.

ATTENTION REVISITED

The brain has an enormous computing capacity, so it is perhaps surprising to realize that we have a very limited ability to multitask. We are generally conscious of our surroundings because inputs from the neuronal circuits for vision, hearing, touch, and other senses enter the thalamus and are then disseminated throughout the brain. However, an awareness of each sensation arises from activity in the ACC. Thus, we are aware of a sound because neurons in the ACC attend to that sensation, thereby giving it priority over the others. The same goes for an awareness of a colored flower or of an object that is being touched. We can flit from one sensation to another very quickly, but we cannot seem to be able to focus on more than one sensation at a time. Pain of course is going to be prioritized over all other sensations because it signifies a threat that might imperil our lives. Consequently, the simple realization above has very important implications for controlling pain because it allows us to propose that *painfulness can be diminished by directing our attention to another sensation.*

There is ample anecdotal evidence that this happens. For example, postsurgical patients exhibited significantly reduced pain when paying attention to music. Thus, the awareness of the music effectively distracted them from the pain. Moreover, the greater the relevance of the distraction, the greater the reduction in painfulness. Attending to an image of a beautiful sunset, or to a religious icon of particular relevance, is a powerful distractor. We know from our own experiences that under certain circumstances we can become oblivious to all sensations, as exemplified by the absent-minded professor who is so focused on an idea that he or she is not aware of what is happening around them. Taken at face value, suffering could be reduced if patients learned to focus their attention elsewhere.

To best implement such a diversionary tactic, we need to know what occurs in the brain during a distraction from pain.[8] In a study designed to answer this question, volunteers were divided into two groups. Both were subjected to a painful stimulus, but one group was distracted during the stimulus and the other was not. The distracted group reported reduced painfulness compared to those in the group who were undistracted. Images of the brain of the distracted group showed increased activity in the PFC and the affective division of the ACC and reduced activity in the thalamus, IC, and cognitive divisions of the ACC (table 12.1). These results are notable because they divided the ACC into two regions—one concerned with painfulness, which was inactivated, and the other with awareness, which showed increased activity due, presumably, to attention being paid to the distraction. In another study, subjects who were asked to focus on a color, followed by heat stimulation, reported that the distraction significantly reduced the intensity of the pain and concurrent fMRI imaging showed significantly greater activation of the periaqueductal gray (PAG).

We cannot directly compare the results of these two studies because they used different procedures, but the conclusion from both is that distraction is a promising way to reduce painfulness and that modules within the pain matrix were involved,

TRAINING ALTERS THE BRAIN

Although it is relatively easy to be distracted from pain, the effect is fleeting and the painfulness soon returns. However, as we shall soon see, there is evidence that the duration of the distraction can be prolonged and, moreover, that it can be willfully controlled. To understand how this is possible, we have to briefly discuss an important phenomenon known as *neuroplasticity*.

Our brains in infancy have many more neurons, connections, and synapses than we have as adults because as we develop, the pathways and circuits used the most are strengthened whereas those that are not used are diminished or lost. The ability to add or remove synapses reflects the fact that the brain is constantly changing with experience, which explains why a nurturing environment in early childhood is especially important. We actively direct some of these changes when we learn to hit a golf ball or do a summersault. In these cases, we are reinforcing circuits to learn a skill, and the more we practice, the more skilled we become and the more the circuits are strengthened. The changes can be very profound, such as when the sense of hearing becomes more acute after blindness. While the ability to alter networks in the brain diminishes, as we get age, older adults are still capable of learning new skills and developing new memories.

Neuroplasticity is almost always discussed as a response to an external event or immediately desired outcome, but suppose it was possible to alter circuits in the brain by turning our own thoughts inward, to train our mind to shape our perceptions and suppress persistent or chronic pain by willfully activating components of the pain matrix, such as those involved in attention? Moreover, just as learning to execute a tennis serve, is it possible with sufficient training that our brains could be rewired to greatly extend the duration of diminished pain? We will devote the remainder of this chapter to build upon what we know to assess whether this is feasible.

COGNITION AND CHRONIC PAIN

The pharmacological approach to treating chronic pain has had only marginal success due to severe side effects from the most effective of the analgesic drugs. In addition, analgesics cost millions of dollars to develop and most require that patients visit a doctor to get a prescription. As a counter to drugs, several nonpharmacological treatments have appeared that are designed to manage chronic pain. Among these, cognitive-based treatment (CBT) strategies have had some success.[9] A central principle underlying CBT is that painfulness is influenced by psychological factors that can be manipulated to reduce the hurt. CBT practices are multifaceted and combine learning how to manage moods, attention, thoughts, and beliefs, coupled with a regime of stretching and various types of physical exercise. Any time a treatment has so many components there will be variations introduced by practitioners, which makes it difficult to attain an accurate assessment of its overall efficacy. Consequently, an effort was made to identify those practices in CBT that appeared to be most efficacious yet simple enough to be taught reproducibly. A primary goal of CBT was to manage *attention* and *belief* and there was increasing evidence that both of these properties of the mind could be controlled by meditation. In light of this new information, some variants of CBT have recently emphasized the mental aspect of pain by incorporating a meditation component. This is particularly meaningful because judging by all of the media attention, meditation alone seems to be a very effective way to reduce pain.

CONTEMPLATION, PAIN, AND SUFFERING

If we wanted to design a way to willfully reduce pain, there is no question that we would want to activate the opioidergic neurons in the PAG. We know that there are connections between the PFC and the PAG and that the placebo effect utilizes the PAG to reduce pain. However, it has not been possible to activate this specific pathway willfully. Another option, however, is to use meditation to divert attention away from the suffering. Suffering is a sensation, and we now know that all sensations flow from the thalamus to the ACC where interactions with the PC and IC determine to which sensation we pay attention. Input from a lesion or injury receives the highest priority because it must be attended to as a possible threat to survival. Given that the more we focus on a sensation, the more everything else recedes, a guiding principle in meditation is that we should be able to make persistent or chronic pain diminish in importance by diverting our attention from the suffering to another sensation or thought.[10] Not to be overlooked is that meditative practices include those that aim to relieve the stress and anxiety that worsen pain.

The idea that meditation can train our minds to control perceptions and emotions has existed at the very fringes of psychology and behaviorism. In the 1960s, we had the transcendental meditation movement and others that were not accepted by the general population for a variety of reasons, foremost among them the skepticism of the medical and scientific communities. Important also was the introduction of drugs like Prozac, which were widely used to reduce the tension and anxiety of modern life. However, an acceptance of meditation practices to willfully control attention has recently gained traction due to the confluence of two very different ideas about suffering.

Suffering in early Western cultures was considered to be an ineffable process that was simply not quantifiable and therefore not a subject for scientific inquiry. Consequently, up until the twentieth century or so, studies of painfulness were consigned to the realm of philosophy and theology. With the advent of neuroanatomy and then neuroscience in the twentieth and twenty-first centuries, it became obvious to most that the suffering emerged from processes in the brain. Even though this had been proposed by Alcmaeon thousands of years earlier, the complexity of the brain still precluded experimental approaches to understanding the source of suffering, and it was largely left to disciplines such as psychology and psychiatry. As we have already documented, this changed relatively recently due to significant advances in cell and molecular neurobiology and the ability of real-time imaging to provide snapshots of networks in the brain that were associated with painfulness. The result was the promise of novel nonpharmacological opportunities to control pain.

Eastern cultures took a very different approach to suffering by using contemplation to understand how the mind could be controlled—that is, they turned inward and developed ways to use the mind to understand the mind. Contemplation in this context means attempting to reveal or clarify the relationship between the mind, emotions, and the world around us. These methods were refined over thousands of years and culminated in what we broadly call meditation. The great religious teacher Buddha was an ardent explorer of the mind who used mediation to gain an understanding of suffering and how it could be reduced with practice. His teachings formed the basis of the Buddhist tradition, which has been perpetuated by the adherents of Zen meditation and other related derivatives. It is quite remarkable to think how much the Eastern practitioners learned about the working of the mind without knowing about neurons, circuits, or networks.

Meditation was not designed specifically to relieve suffering but rather to quiet the mind by achieving a greater knowledge of self and control over our memories and emotions. Quieting the mind naturally relieves stress, which we know can lessen pain. Each of us has an innate basal meditative state that we are in when our mind is free of external influences. This level of withdrawal differs from individual to individual and is associated with pain tolerance. Thus, the deeper your natural meditative state, the more resistant you are to a painful event. This partially explains why painfulness is subjective but also suggests that manipulating the basal state might be a way to alter the experience of pain. Yoga exercises are designed to help practitioners enter a meditative state by focusing on breathing, body movements, and positions as a way of focusing on a physical activity and away from extraneous influences. Runners can sometimes achieve the same state by synchronizing their breathing with their pacing, thereby taking attention away from pain.

MINDFULNESS

There are several forms of Buddhist meditation, and the most successful ones incorporate the Eastern practice of mindfulness. For thousands of years Buddhist monks have claimed that mindfulness meditation significantly alters the subjective experience of pain. It is said that they fully experience the sensory aspect of pain, but after an evaluation, the pain is dismissed. There are also many stories in Eastern lore about how mindfulness practitioners can perform extraordinary feats by exerting control over their mental processes. Evidently there is a kernel of truth in these claims because recent studies have shown that mindfulness-based therapies can successfully relieve suffering and that even short-term mindfulness training can have a significant analgesic effect.[11] Briefly, mindfulness is a mental state that is achieved by focusing awareness on thoughts as they flow through the mind. At the basic level, each thought is not judged but is allowed to merely dissipate without effect. In common with most forms of meditation, a primary goal is to attain a calm mind and a minimal or basal state of awareness. On another, more advanced level, mindfulness purposely and willfully directs the mind to attend to a particular sensation and away from a thought or sensation that is to be avoided. The purpose here is to skew this unwanted intrusion to irrelevance. Among the early successes using this approach was a program in mindfulness-based stress reduction (MBSR) that was designed in the late 1970s by Jon Kabat-Zinn and colleagues at the University of Massachusetts Medical School.[12] Many practitioners of MBSR learned how to reduce their chronic pain and improve their quality of life, and the course is now offered at many locations throughout the United States. When added to the reported success achieved using cognitive-based treatments, we can begin to see the potential power in training the mind to overcome pain. Nevertheless, questions remain.

Although mindfulness meditation is effective, there is still no understanding as to why or how it works. Many books offer various theories about the underlying mechanisms yet no verifiable explanation has been attained. This is largely because most scholars of meditation do not recognize the mind as a property of the brain. Why this should be is certainly not clear. Many studies of injuries in the eighteenth and nineteenth centuries clearly linked focal brain damage to loss of a discrete function. Speech, for example, arises from the activity of a discrete group of neurons in the left frontal lobe.[13] More recent studies of direct brain stimulation, some of which we have mentioned, elicit complex sensations and emotions that provide direct evidence that these properties emerge from the substance of the brain and not from some external spirit or force. Consequently, to argue that consciousness and awareness emerge from an as yet to be defined extracorporeal force makes little sense. Some of these explanations are so convoluted that they lead to questioning the very meaning of what it is to exist. We are taking an agnostic position and accept that meditative introspection works to relieve pain by evoking processes that cannot be explained by current science. This is the same as our acknowledging that sensations emerge from neuronal circuits in the brain by processes that are likewise not understood. Nevertheless, by melding the Eastern and Western traditions we can obtain insights into how meditation reduces suffering and how it works within the bounds of what we know about the workings of the pain matrix.

The practice of mindfulness can be roughly apportioned into two disciplines that differ in how they process sensations, thoughts, and emotions.[14] Both require the mastery of certain skills. The first discipline uses *focused attention* (*samatha* in the Pāli language), which promotes a detachment from external and internal events. During focused attention, the practitioner focuses on a mantra, a word that is repeated over and over, or on a simple function such as breathing. Breathing is particularly effective because inhaling deeply and exhaling slowly alone reduces tension and anxiety by ensuring centers in the brain an adequate supply of oxygen. When a distracting sensation or thought intrudes, it is dismissed and the focus returns to the breathing. This is not a trivial accomplishment because the mind is constantly in a state of flux in which sensations, thoughts, and even memories compete for attention. Nevertheless, as training progresses the practitioner gradually learns to sustain attention on the breathing, and the nonjudgmental awareness of sensations, such as painfulness, becomes fleeting, The ultimate goal is to achieve a state of relaxation in which the mind is focused on itself at the exclusion of extraneous fear or stress.

The other discipline within mindfulness meditation employs what is known as *open monitoring* (*vipassana*), where any thoughts or experiences, such as painfulness, that intrude during focused attention are evaluated and then dismissed as not important. Put another way, open monitoring uncouples an awareness of the pain from the painfulness, which is reminiscent of the response of the lobotomy patient to his severe injury. In open monitoring, however, the result emerges from a willful decision that the unpleasantness is not important.

We mentioned that unequivocally assessing the efficacy of meditation is difficult when there are differences in how meditation is taught. Some instructors tell students to close their eyes, which eliminates the transmission of visual information from the thalamus to the ACC. Others allow students to recline comfortably during meditation, whereas others require a sitting position, which can be painful. In fact, many yogis would meditate for many hours and they had the very real goal of having to dismiss the pain as it occurred in their joints and muscles. We believe that for the typical practitioner of mindfulness meditation, sitting or lying down, or having the eyes open or closed, will not have a significant difference in the outcome provided the practitioner is comfortable.

There is a subtle but important difference between focused attention and open monitoring with regard to the role of the pain matrix. Focused attention is a relatively simple process that can reduce pain by attending to breathing while diverting painfulness to another corner of the mind. Breathing is essentially neutral, and the inputs to the PFC and IC from cortical sites that deal with higher cognition, along with the outputs to centers for attention, will both be quiescent. In contrast, open monitoring actively acknowledges the painfulness but decides that it is not significant. Since all decisions involve knowledge and belief, which in turn depend on interactions with the reward system, we would expect open monitoring to increase activity in many of the modules in the pain matrix, especially the PFC and IC. Consequently, the activity in the PFC and IC would be much more extensive than what occurs during focused attention and would be similar to what occurs during hypnosis and successful treatment with a placebo. These areas and others would also contribute to an acceptance of the pain; we mentioned the significance of acceptance in our discussion of martyrdom and the reward system.

MINDFULNESS AND BRAIN WAVES

To assess the effectiveness of any treatment requires an objective measure of success. A primary benefit of mindfulness is the calming effect that reduces the stress and anxiety that exacerbates pain; it is actually possible to monitor this state of mind by making use of technology that detects the waves of electrical activity that radiate from the brain. The waves arise when extensive arrays of neurons in the brain generate action potentials in relative synchrony so that they can be detected by attaching electrodes to the scalp. These waves were first detected in 1924 by the German physiologist and psychiatrist Hans Berger using an electroencephalogram (EEG), which he developed.[15] We have not discussed this way of monitoring brain activity because it has been largely superseded by the far more informative imaging techniques. Nevertheless, brain waves are useful because they measure activity that correlates with levels of consciousness. Several different waves can be distinguished by the frequency of the oscillations. Alpha waves have the lowest frequency and are present when the brain is in a relaxed state, such when you're daydreaming. When alpha oscillations are prominent, sensory inputs are minimized, the mind is generally clear of unwanted thoughts, and there is a marked reduction in anxiety.

Significantly, mindfulness training tends to produce noticeably more alpha waves. Thus, the EEG, which objectively measures brain activity, corroborates the considerable anecdotal evidence on the ability of mindfulness to quiet the brain.

As might be predicted, when our brain shifts gears to willfully dwell on a specific thought, as occurs during open monitoring, alpha oscillations tend to be replaced by the higher frequency gamma waves. These waves reflect simultaneous processing of information from different brain areas and are associated with higher states of conscious perception. Gamma waves predominate when our minds are actively learning or in hyperactivity mode. Uncontrolled gamma waves can cause anxiety, so it was surprising that Tibetan monks in a meditative state had levels of gamma waves that were two to three times higher than the resting level. Their minds were in an extraordinary level of alertness even though they were in a relaxed state of meditation. An explanation for this apparent contradiction is that the gamma waves were manifestations of the mind focusing on itself.

Most of the original testament as to the efficacy of mindfulness indicated that it can reduce the suffering from chronic pain. Many of these assessments were based on self-reporting—the person being evaluated determined the intensity of the painfulness, which introduced an element of subjectivity. On the other hand, the alpha wave studies cited provided a more objective measure that corroborated the reporting from the patients, at least with regard to relieving stress. However, one very important possibility should be considered, which is that mindfulness relieves pain via a placebo effect. Moreover, solid support for the effectiveness of mindfulness to relieve pain would be to show that the meditative state engages components of the pain matrix in ways that would be predicted based on what we know about their roles in suffering. To address both of these important issues, let's compare the activity of the various modules of the pain matrix in patients successfully being treated with a placebo with subjects during mindfulness meditation.

MINDFULNESS REAL-TIME IMAGING

fMRI imaging of mindfulness-focused attention showed activation of areas that we would have predicted from what we have learned.[16] Thus, there was activation of the nose and throat region of the somatosensory cortex (i.e., the sensory homunculus) that was due to the focus on breathing and in the ACC that we know is involved in awareness. This was to be expected. The crucial experiment was to compare the brain scans obtained when a painful heat stimulus was applied to the lower leg of subjects who had mindfulness training versus those who did not. The heat activates the TRPV1 receptors on the terminals of the C-type nociceptive neurons.

As anticipated, images of the nonmeditating group showed increased activity in the somatosensory cortex in the postcentral gyrus (PCG) in the region of the lower leg and in the other regions in the pain neuromatrix involved in processing the effects of an injury (table 12.2). Thus, they were experiencing pain as usual. Strikingly, the meditating group showed *reduced* activity in the leg region of the sensory cortex, as well as in the thalamus, amygdala, and the periaqueductal gray. Thus, meditation appears to attenuate pain by blocking the nociceptive pathway at the level of the thalamus. Moreover, the imaging also revealed an increase in the activity of neurons in the PFC in the meditating group. In another similar study, it was shown that meditation increased brain activity in areas of the ACC and IC.

TABLE 12.2 CHANGES IN THE ACTIVITY OF COMPONENTS IN THE PAIN MATRIX IN RESPONSE TO AN INJURY, PLACEBO, AND MEDITATION

	Injury	Placebo	Meditation
PCG	increase	decrease	decrease
Thalamus	increase	decrease	decrease
ACC	increase	decrease	increase
IC	increase	decrease	increase
PFC	increase	increase	increase
PAG	increase*	increase	decrease
Amygdala	increase**	decrease	decrease
N. accumbens		increase	

*After extreme injury or stress. **When fear is involved. See text for details.

Table 12.2 shows the similarities and differences in the responses between the meditating practitioners and the placebo patients, and it is gratifying that most can be explained by what we know about the function of the modules within the pain matrix. Thus, there was a decrease in the activity in both the IC and ACC in the placebo patients. Since activity in the IC-ACC network is essential for the expression of pain, this explains why a placebo can reduce painfulness. Remember that imaging studies of chronic pain patients showed enhanced activity in the IC and ACC. On the other hand, the increased activity in the ACC in the meditating group would be expected because they are focusing on an awareness of thoughts emanating from the increased activity in the PFC. But why is the PFC active during meditation? After all, the reduced activity in the thalamus means that there will be diminished input from vision, hearing, and painfulness. A reasonable explanation is that circuits in the PFC are active because they are evaluating thoughts that originate within the brain, i.e., from other regions of the cerebral cortex. This is in accord with the vast increase in gamma wave recordings from the meditating monks mentioned earlier. There was also increased activity in the PFC in the placebo group, which we would expect given its role in the belief and knowledge that are so important to its success. Lastly, both the placebo patients and meditating practitioners exhibited reduced activity in the amygdala, indicating a reduction in fear. This is in stark contrast to images from chronic pain patients who show increased amygdala activity. As we can see, there is a good correspondence between what we have learned about the functions of the components in the pain matrix and the reductions in painfulness due to a placebo or mindfulness.

The one truly unexpected revelation from the comparison was the reduced activity in the PAG in the meditation group and its activation in the placebo patients (table 12.2).[17] Consequently, mindfulness meditation diminishes the experience of pain via pathways that do not involve the opioid system. This distinction is supported by studies showing that naloxone does not attenuate the reduction of pain seen during mediation but, as we mentioned in an earlier chapter, it does so with the placebo. We can therefore conclude that placebos and meditation use different components of the pain matrix to relieve pain.

Taken together, these findings are important because they demonstrate that the neural mechanisms involved in mindfulness-based pain relief are consistent with the control of modules in the pain matrix that are involved in cognitive control of awareness and painfulness. The placebo effect is equally capable of relieving painfulness but alters the activity of different components in the pain matrix. Thus, pain can be modulated at not just one place in the brain but at multiple processing sites in the matrix.

SELF-REGULATING PAIN

We are aware of a lesion via connections between the thalamus and the ACC, but the painfulness is imposed by inputs to the ACC. Judging by the results from meditation and the placebo, the most important of these is the connection between the IC and the ACC. But remember that the IC receives inputs from the PFC and other areas of the cortex that mitigate the suffering based on belief, knowledge, etc. Now suppose it were possible to learn how to activate these areas of the cortex directly without having to meditate or depend on a placebo? One way would be to use a feedback system that would guide the patient in learning how to control a particular physiological or behavioral outcome. Courses in biofeedback training have helped patients learn how to regulate their heartbeat and other autonomic functions, and their success has resulted in the development of neurofeedback programs. These have been designed to teach patients how to self-regulate brain functions so as to modulate the electrical activity of areas of the brain involved in experiencing pain. One such course uses an EEG to monitor brain waves during a prescribed activity. We have already discussed how the presence of alpha waves correlates directly with the mindfulness mental state of relaxation, which reduces the stress that exacerbates pain. It turns out that subjects can be trained to enter this relaxed state by willfully activating mental processes that maximize the appearance of alpha waves. This has had some success in lessening the experience of pain and has the decided advantage of reducing the time and effort required to attain the relaxed state via meditation.

The neural mechanisms responsible for the EEG rhythm are widespread and therefore span multiple modules within the pain matrix. More useful would be to selectively target only the networks that are involved in suffering. Toward this end, very promising results were obtained in a collaborative effort by scientists from Stanford and MIT-Harvard.[18] They targeted the ACC as an essential component in the pain system and used real-time fMRI to teach a group of subjects how to willfully upregulate and downregulate the activity of neurons in the ACC. There was a cognitive component to the training because the subjects were told that they would be attempting to alternately increase and then decrease activation in the target brain region and that the fMRI scan would provide real-time feedback as to their success. In other words, the subjects were motivated to achieve a goal. Subjects who could successfully regulate the activity of the ACC were then given a noxious, localized thermal stimulus. When asked to evaluate the pain, they reported that the painfulness was increased when they deliberately increased the activity of the ACC and that the painfulness diminished when they reduced the activity in the ACC. This striking result had obvious implications for managing pain. The results also demonstrated directly that the ACC regulates the degree of suffering. Three other groups were used to control for a placebo and nonspecific effects.

The definitive experiment was then carried out on a small group of chronic pain patients. Using fMRI to provide feedback as before, the patients were trained to control the activity in the ACC. Unlike the subjects above, the patients were not given the noxious thermal stimulus but were instead asked to evaluate their ongoing pain. The results were the same: patients experienced a significant decrease in their level of painfulness when they deliberately reduced activity in the ACC. Although more studies with larger numbers of patients are clearly required, the results show that individuals can gain voluntary control over the activation of a specific brain region and that the control over the ACC was powerful enough to impact chronic clinical pain. An interesting practical issue is to determine how long the patients are able to control the ACC and pain. In other words, does the training cause a rewiring of the brain, as occurs when we are trained to carry out a specific physical activity?

• • •

We first showed that the activity of modules within the pain matrix is responsible for the experience of pain. The somatosensory component sends signals that provide an awareness of an injury or inflammation, but the ultimate extent of the painfulness is determined by both the affective and cognitive components of the matrix. The affective systems contribute to fear, based on previous experiences, and reward, which takes into consideration belief and value, whereas the cognitive components control the outcome based on knowledge and memories. This is a science-based and relatively mechanistic view of how pain arises. A very different approach to understanding pain is to use the mind to control suffering as exemplified by cognitive-based therapies and mindfulness meditation. What has emerged is that our understanding of the pain matrix validates the outcomes obtained by the mindfulness, and vice versa. The goal is now to use what we have learned to open a new and promising stage in pain management, which will be discussed in the final chapter.

13

PAIN MANAGEMENT

Present and Future

We have discussed pain and suffering from many vantage points, and our next goal is to use what we have learned to make informed decisions about how to best treat pain, both now and in the future. The goal is to alleviate all forms of chronic pain, but in cases where the pain is intractable, we want to reduce the suffering to manageable levels so that patients can lead productive lives. We initially focused on the enzymes, channels, and receptors in the somatosensory system that are responsible for the initial perception of pain. The search for additional factors is continuing, which is important because it provides the pharmaceutical industry with targets in the quest to formulate new analgesics. In subsequent chapters, we learned that pain is not just another perception of the external world but a highly complex sensation that is modulated in ways that would have been unfathomable just fifty years ago. Thus, we introduced the pathways in the brain that descend to the spinal cord to regulate the nociceptive pathway via the release of opioids and other neurotransmitters. Most significant, however, was the realization that pain and suffering are actually *experienced* and that the somatosensory system is merely one component of a network of discrete centers within the brain that we depicted in the pain matrix. This understanding greatly broadened the view of pain because the matrix included centers that impose controls on the experience of pain and, for the first time, provided an explanation for pain from bereavement and other psychological causes. Most important, it resulted in studies indicating that suffering can be willfully controlled by manipulating components of the matrix. Thus, we have two approaches to managing pain—one rooted in pharmacology and the other based on mindfulness—and we now must consider the merits of each as we progress toward our goal.

THE PRESENT

The Pharmacological Approach

MAKING OPIATES SAFE

Chapter 9 described the challenges and pitfalls of drug development and the need for clarity in selecting the best target. Nevertheless, there is still a real need for pharmacological interventions. We have known for centuries that opium is an effective analgesic; its refined derivatives, such as morphine and oxycodone, are currently the primary drugs for intractable pain and they will likely continue to be so into the foreseeable future. Opiates are extraordinarily effective analgesics for many types of chronic pain and administering an opiate as a pill or by injection is simple and relatively inexpensive. The downside is the risks posed by the very serious side effects. As a practical matter, limiting the more extreme side effects would vastly increase the utility of the opiates. We now have evidence that the most serious risk, addiction, is caused by irrational activation of the reward system. It appears that the pleasure derived from taking a narcotic overrides consideration of the negative consequences and reinforces the motivation to continue. Consequently, one avenue for drug development would be to block the activation of the reward neurons in the nucleus accumbens. This would require the identification of the key molecules in the neurons that are involved in the activation: a very difficult, but not impossible, task.

A better approach would be to target the phenomenon known as *tolerance*, which is an alteration in brain chemistry such that it requires a larger dose of an opiate to achieve the desired level of pain relief. Opiates mimic the endogenous endorphins (opioids) by binding to receptors in the spinal cord. Since these receptors are also present in the respiratory system and brain, as the dosage is increased, so is the severity of the side effects, most notably the potential for respiratory distress. The higher doses also increase the dependence on the drug and the intensity of the withdrawal symptoms. The cumulative effect is a craving for the drug and the likelihood of developing an addiction. Thus, eliminating the tolerance should retain the analgesic properties and remove many of the side effects. Fortunately, efforts are underway to characterize the molecules involved in the development of the tolerance, which would clearly be a tremendous advance in pain management.

Stress-induced analgesia, which provides the most profound relief from pain, is triggered when an extremely severe injury activates the collateral nociceptive pathway to the opioidergic neurons in the periaqueductal gray (PAG). The release of the opioids from their terminals at the synapse between the first- and second-order C-type neurons prevents both the awareness and the suffering from the injury. The effectiveness of this descending system is evident from the power of opiates to relieve pain. However, gamma-amino-butyric acid (GABA), serotonin, and noradrenalin are also released by these descending systems. Although the drugs that influence the levels of these compounds were designed for other purposes, some, such as Valium, pregabalin (Lyrica), and the reuptake inhibitors Prozac and Reboxetine, mitigate the severity of the pain by lifting depression and lowering anxiety. Gabapentin, which blocks voltage-dependent calcium channels, has been approved for treating shingles and diabetic neuropathy. All of these drugs should be considered important adjunct treatments for chronic pain.

MEDICAL MARIJUANA

From one perspective, the availability of opiates has actually been a distraction because they have reduced efforts to find other effective analgesics. This could now change due to the increased interest in marijuana as a source of analgesia. Scientists have already shown the antinociceptive effects of the anandamide-CB1 receptor and the anandamide-CB2 receptor systems. As discussed in chapter 9, the CB1 receptor is distributed throughout the brain and is responsible for many of the unwanted psychogenic responses due to THC. Consequently, this system does not appear promising for further development. Much better is the analgesia triggered by the CB2 receptor. This occurs in the periphery and is due to inhibiting the activation of cells in the immune system and by desensitizing the TRPV1 channel. We know that many types of chronic pain have an inflammatory component, and drugs that target the CB2 receptor system have already proven effective for treatments of inflammatory and neuropathic pain. The most promising ingredient in marijuana for relieving pain, however, is cannabidiol (CBD). This compound blocks the transmission of nociceptive information in a variety of ways, some of which are unique. Cannabidiol-induced analgesia is not accompanied by serious side effects and drugs containing CBD are being marketed for chronic neuropathic pain. Much more research is needed to determine exactly how the CB2 receptor system and cannabidiol relieve pain and we are confident that as the details unfold, many new targets for drug development will be identified. Until then, medical marijuana, as it is called, is a useful adjunct in pain management.

THE SOMATOSENSORY SYSTEM: NEW TARGETS

We have discussed the key molecular components responsible for the allodynia and hyperalgesia that are concomitants of chronic pain. Unfortunately, attempts to develop analgesics against the more promising of these, such as the TRPV1 receptor/channel or the NMDA receptor, have failed due to the appearance of unacceptable side effects. Many of these effects occur because administering a drug in pill form, or via intravenous injection, results in the drug being distributed throughout the body where it can interfere with the function of the target in other organs. Availability is another problem. Methadone is probably the most effective of the NMDA inhibitors, but its usefulness is undercut because it is retained in the body for many hours and therefore has ample time to do damage. The solution is direct delivery of the drug to the target area, which is an area of active research. Important also is selectivity. The NaV1.7, NaV1.8, and NaV1.9 sodium channels all appear to have some role in chronic inflammatory pain and are therefore promising targets because they are localized primarily to the C-type nociceptive neurons; blocking their activity should be very effective against nociceptive pain.[1] The challenge is that a drug that blocks these channels would have to be very highly selective because any overlap with sodium channels elsewhere would interfere with the generation of action potentials in many other classes of neurons. Selectivity is always a major problem in drug development, especially with regard to analgesics.[2] Improvements in drug design based on detailed structural information can potentially overcome this selectivity problem by focusing on differences between the various forms of the target.

Other obvious targets would be the agents responsible for the inflammatory responses because these can persist for long periods and often precede some forms of chronic pain. Several analgesics block components of the inflammatory cascade, such as the Cox inhibitors, which prevent the synthesis of agents that activate the terminals of nociceptive neuron. The Roche drug Actemra blocks the synthesis of interleukin-6 (IL-6), and it should be beneficial because we now know that IL-6 contributes to the anxiety that enhances the experience of pain.

Of course, what distinguishes chronic pain from other forms of pain is its duration. Consequently, a reasonable approach to prevent chronic pain would be to block components in the nociceptive pathway to prevent the transition of acute pain to the chronic. The focus here would be on the molecular events that extend the duration of the pain experience and we know of two: the late phase of long-term potentiation (LTP) and long-term hyperexcitability (LTH). Remember that LTP extends the duration pain by sensitizing the postsynaptic terminal of the second-order neurons. The increased sensitivity (allodynia) means that even a few action potentials from a light touch at the lesion site (or elsewhere) will generate multiple action potentials in the second-order neurons that will travel to the brain. Selectively blocking LTP would decrease the sensitivity and should leave normal transmission intact. As discussed at length in earlier chapters, we have a good understanding of the molecular events responsible for LTP and we already discussed the NMDA receptor/channel and several other compounds of interest. Access is an obstacle, however, because LTP occurs at synapses within the spinal cord, which is protected by the blood–brain barriers (BBB). In addition, the late phase of LTP might contribute to persistent pain but does not last long enough to be responsible for chronic pain. The focus then shifts to the long term hyperexcitability (LTH) that is already known to be associated with several chronic pain conditions.

LTH results from a phenotypic change, meaning that it can potentially last indefinitely and could extend the duration of the late-phase LTP.[3] The induction of LTH depends on events that occur in the cell bodies of the C-type nociceptive neurons in peripheral ganglia. A big advantage here is these neurons are not protected by a BBB and are directly accessible to drugs in the circulatory system. Studies have established that an essential factor in the appearance of LTH is the activation of protein kinase G (PKG); a very potent and selective inhibitor of PKG that reduces various forms of chronic pain in animal models has already been synthesized.[4] Notable was that the inhibitor was synthesized by rational design in which computer models of the active site in PKG were used to guide the development of the most effective inhibitor. Consequently, it only required the synthesis of 150 compounds to obtain a very potent and highly selective inhibitor of PKG. This method is far removed from the typical pharmacological approach, which is to synthesize thousands of potential agents at great expense. The inhibitor has great promise, but it is still in the discovery phase and has many hurdles to clear before it will be ready for clinical trials.

PKG is at the beginning of a cascade of events that ultimately results in the change responsible for the increase in excitability. There are likely to be several components in the cascade, any one of which would also be a suitable target. How long LTH lasts is not known and there might be a finite window before the painfulness no longer depends on signals from the periphery. Nevertheless, a drug that blocks LTH would still be useful for treating fibromyalgia or neuropathic pain if it was administered before the transition to chronic pain.

Another promising target is nerve growth factor (NGF). NGF levels are elevated in preclinical models of both inflammation and peripheral nerve injury, and the concentration of NGF is increased in chronic pain conditions such as interstitial cystitis, prostatitis, arthritis, chronic headaches, cancer pain, and some forms of neuropathy.[5] We discussed how NGF acts in concert with other components at peripheral terminals to initiate action potentials and also how, after a lesion, NGF is retrogradely transported to cell bodies where it promotes the production of important proteins, such as sodium channels. NGF therefore has an important role both in the initiation of pain and in prolonging pain. Pfizer developed a novel NGF inhibitor, Tanezumab, especially for the treatment of the pain from osteoarthritis.[6] During clinical trials Tanezumab successfully relieved the pain relative to a placebo, but the side effects were deemed unacceptable.[7] Work on developing NGF inhibitors is continuing and updates on progress appear online.

As studies of the molecular changes that contribute to chronic pain continue, more targets will undoubtedly be identified, so there is still hope that a nonaddictive pharmacologic analgesic for chronic pain will eventually become available. In the interim, there is increasing evidence that pain can be effectively mitigated using nonpharmacological approaches.

The overriding problem in pain management is that even opiates do not effectively alleviate all types of chronic pain because suffering can have psychological as well as physical causes. This means that inhibiting a single enzyme, channel, or receptor in the somatosensory system will not reduce all forms of chronic pain. However, since all painfulness arises from the activity of neurons in specific modules within the pain matrix, inhibiting this activity will block pain from all sources, *even after the pain is established.* We emphasize the last point because some forms of chronic pain such as bereavement are perpetuated by activities that are independent of any external inputs. Neuroscientists are only beginning to characterize the molecular components responsible for these activities, but these efforts are not essential because we don't have to target a particular molecule, just an outcome. In other words, it should be possible to mitigate the experience of pain by manipulating the function of certain modules in the matrix.

COGNITIVE-BASED APPROACHES: MINDFULNESS

We have documented the ability of belief, knowledge, and reward to ameliorate pain by controlling the activity of certain modules within the pain matrix and have emphasized the importance of attention in the process. In essence, we can learn to use the mind to distract our attention from the pain, as occurs via hypnosis, a placebo, or meditation. The drawback to hypnosis is that only a small percentage of the population can attain the deep hypnotic state needed to alleviate pain. A placebo can reach a larger population but usually requires a long-term patient–doctor relationship because the outcome depends on trust. As of this writing, mindfulness-based meditation is the best nonpharmacological treatment for chronic pain because it can benefit the greatest number of patients, is low risk, and is inexpensive, although it does require training.

There are two states of mindfulness that can reduce pain, and each traditionally requires the guidance of a qualified teacher. Attaining the lowest state, focused attention, is relatively easy and once mastered, the practitioner can enter this state at will without much preparation. This state calms the mind and directly benefits those in pain in two ways. First, it reduces any fear due to the activation of the amygdala, since imaging studies have shown that an active amygdala contributes to the pain of chronic fibromyalgia. This is certainly understandable because fibromyalgia pain is not constant—it can unexpectedly "flare up" resulting in fear of an impending episode. Removing the fear also improves mood and the quality of life. Second, focused attention blocks the stress-induced synthesis of inflammatory factors whose presence exacerbates pain. Having the patience to commit to learning how to meditate effectively might be difficult for some of those in pain and they could take advantage of the findings that the time required for training can be reduced by using biofeedback to enhance the production of alpha brain waves that correlate directly with the presence of the relaxed state.

The next level, open monitoring, aims to achieve a state of mindfulness in which the practitioner willfully diverts attention away from the pain. Thousands of years of practice have provided anecdotal support to the claims that attaining this level of mindfulness can significantly reduce ongoing pain. These claims have now been corroborated, and we can state with reasonable certainty that skilled practitioners can willfully modulate the activity of components within the pain matrix in ways that are consistent with analgesia. Moreover, by manipulating centralized components of the matrix, such as the ACC, mindfulness can attenuate both physical and psychological pain. The primary obstacle is that achieving this level of competence requires time, patience, and dedication. If we look to the future, we can envision how this obstacle can be surmounted by technology.

THE FUTURE

Electronically Induced Analgesia

The essence of mindfulness training is learning how to manipulate the function of modules in the brain. One could argue that since mindfulness meditation works to relieve painfulness, what was actually achieved by identifying all of the modules? The answer is that identifying the modules gives us the opportunity to manipulate their activity without meditation.

We have shown that meditation can alter the function of the modules, but this overlooks the fact that what is actually being manipulated is the *activity* of the neurons in these modules, and we know that this activity is actually a manifestation of action potentials and synaptic transmission. We could therefore eliminate the training by devising ways to externally regulate the essential activity. One possibility is to use a procedure known as deep brain stimulation (DBS), which can control the excitability of the neurons in a module via an electrode implanted into the brain. DBS has been used for decades with mixed results.[8] An early study that stimulated the periaqueductal gray (PAG) provided relief for some patients but not others. Stimulation of the prefrontal cortex (PFC) also worked in some patients. The failures in these early studies could be explained if the placement of the electrode was imprecise and did not stimulate the appropriate neurons. Somewhat more successful were studies using DBS to treat the stroke victims who develop Dejerine–Roussy syndrome, in which damage to the thalamus results in neuropathic pain.[9] Many of these patients exhibit severe allodynia that is triggered by a stimulus as mild as a mere touch. A particularly terrible form of this syndrome occurs when the pain arises because thalamic neurons become spontaneously active. This form of central pain is very refractory to treatments. DBS stimulation of the thalamus provided pain relief in some patients that lasted as long as a year. More recent studies using DBS were more successful in alleviating pain in several chronic conditions, most notably peripheral neuropathies.

In 1963, in Spain, a man entered a bull ring holding a small box. He was noticed by the bull, which immediately charged, but when the bull got close, the man pressed a button on the box and the bull immediately stopped and placidly walked away. The man, the physiologist José Manuel Rodríguez Delgado, had just demonstrated that he could regulate the bull's behavior via stimulation of electrodes that he had implanted into the bull's brain. This feat was hailed around the world but was criticized by scientists as being a mere stunt. But what a spectacular one! Delgado was a professor at Yale University and during the middle decades of the twentieth century, he showed how electricity could be used to elicit rage, anxiety, pleasure, drowsiness, and involuntary movements in animal and human subjects.[10] There were objections to his work on humans as being unethical because he could make his subjects perform simple actions against their will by stimulating regions of the brain. The criticisms took hold, and his studies on the alteration of behavior by brain stimulation were halted. Nevertheless, his experiments and the studies done provided clear "proof of principle" that deep brain stimulation can be used to regulate the activity of neurons in the brain.

There have been marked improvements in the design of interfaces between the brain and external electronic devices. For example, electrodes implanted into a motor region in the brain of paralyzed patients was programmed to allow the patient to control switches by thinking about activating certain motor functions. Advances in the miniaturization of wires have become important because the finer the wire, the less damage it will do to brain tissue. The resolution of brain imaging also continues to be refined; many of the areas delineated on Brodmann's map have been further divided into even smaller regions with more specific functions. This level of detail means that an electrode can be more precisely placed within a module to regulate only the neurons involved in pain with less collateral stimulation. Finally, a massive worldwide project has focused on defining the interconnections between all of the regions of the brain that will further increase our understanding of how the various modules of the pain matrix interact with each other and with neurons throughout the brain.

Analgesia by Light: Optogenetics

A remarkable procedure has been developed by which the activity of neurons in the brain can be manipulated using pulses of light. The process, known as optogenetics, involves using genetically engineered neurons to express light-sensitive proteins called opsins.[11] Opsins are naturally occurring transmembrane proteins, each of which can be activated by specific and narrow bandwidths of light (fig. 13.1). Some opsins are light-gated ion channels that open or close when the neuron is exposed to light of the correct frequency. Others have been designed to control a specific intracellular pathway. Since neurons can be engineered to express more than opsin type, the electrophysiological properties of the neurons can be regulated selectively by activating the various opsins using pulses of light. It is really quite amazing!

Activation of inserted sodium (Na^+) channels, for example, will initiate action potentials, whereas activation of inserted chloride (Cl^-) channels will reduce the generation of action potentials.

Opsins can be expressed in subsets of neurons using viral vectors that carry the opsin gene. The vectors are injected into the brain region of interest and the neurons that take up the virus will then express the opsin protein. The activity of the neurons

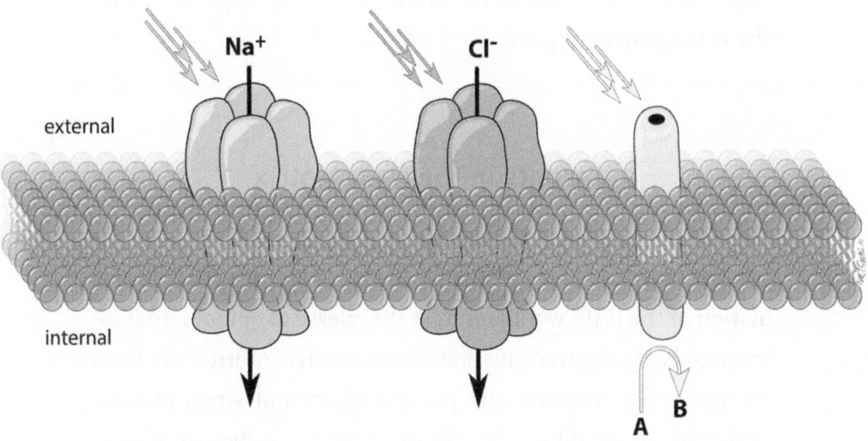

FIGURE 13.1 Schematic representation of three different types of opsins embedded in a neuronal membrane. The two opsins on the left are light-gated channels that open in response to incident laser light of specific frequency (arrows). Activation of the far-right structure results in the activation of an enzyme cascade within the neuron. Insertion of more than one opsin type into a subset of neurons allows rapid control of their electrical activity and enzymatic activity.

can then be controlled by delivering pulses of laser light through very fine (200 uM in diameter) optic fibers inserted into the area. Since lasers deliver a narrow bandwidth of light, it is possible to selectively control the activity of each opsin separately so that the activity of a selected group of neurons can be regulated by merely delivering light of the appropriate frequency. Optogenetics is revolutionizing the field of neuroscience and has already been successfully employed to study the neural circuits underlying mood disorders, addiction, and Alzheimer's disease, among others. Although there are still difficulties to overcome, optogenetics is superior to DBS because it is less invasive and offers much finer spatial and temporal control over the activity of the target neurons. It should therefore have many applications for the management of pain.

Targets in the Pain Matrix

Which module presents the best target to regulate pain? At present, there are two candidates: the PAG and the ACC. Stimulation of the PAG would result in the release of opiods and serotonin to block the transmission of nociceptive information from the body. This explains the placebo effect and stress-induced analgesia but would not be effective against psychological pain. In contrast, there is good evidence that the ACC modulates the experiencing of all types of pain. We know that the neurons in the ACC are involved in both awareness and painfulness, are inactivated during hypnosis- and placebo-induced analgesia, and are activated when pain is enhanced by stress, anxiety or anticipation. Remember the lobotomy patient who somehow knew that he had severely burned his hand but did not care? Subsequent studies showed that a similar separation of affect (the painfulness) from the awareness could be achieved by surgically removing the ACC in chronic pain patients who could not obtain relief from any other treatments. Notable was that the patients reported that the pain was present, but less bothersome, replicating the responses from the lobotomy patient. This finding is very significant because it shows that the treatment did not prevent the patients from recognizing that they were injured. The surgery on the ACC was extensive and had considerable risk. As an alternative, a group at Oxford University found that bilateral DBS of the ACC effectively relieved pain from some patients suffering from a variety of conditions.[12] As research progresses, other important circuits for pain will undoubtedly be discovered, but at present the findings support the idea that the ACC is a central nodule in controlling suffering from both the external world via the somatosensory system and the internal world responsible for psychological pain. Especially important were the studies we discussed in the previous chapter, in which patients could be trained to ameliorate their pain by willfully reducing the activity of the ACC.

Based on all we have learned in this book, we can envision a future in which patients can remotely reduce their pain by controlling the activity of the neurons in the ACC (or some other area), either via an implanted electrode or by regulating the activity of the target neurons by pulses of light delivered via optic fibers. There are many permutations, but all of these possibilities emerged from the studies that showed how pain is regulated by the components of the pain matrix.

ACKNOWLEDGMENTS

My sincere thanks go to my good friend and colleague Michael Sivitz, MD, who read an earlier version of the manuscript and offered numerous suggestions and insights that certainly made this book more readable. Thanks also to Professor Charles Noback, a pioneering neuroanatomist at Columbia University, who gently guided me through my initial efforts to understand the complexity of the human nervous system. This book would not have been possible with the support and encouragement I received from Ms. Miranda Martin and her team at Columbia University Press Finally, I must express my gratitude for having had the opportunity to teach so many talented and inquisitive students whose probing questions deepened my understanding of human anatomy and the brain..

NOTES

1. PAIN AS A PROPERTY OF THE NERVOUS SYSTEM

1. There is no more vexing problem in science than trying to understand consciousness. A debate about whether this is even possible is presented in the book by M. Bennett, D. Dennett, P. Hacker, and J. Searle, *Neuroscience and Philosophy: Brain, Mind, and Language* (New York: Columbia University Press, 2007). The arguments for and against the idea of consciousness are presented, and there are discussions as to whether there are sufficient words in English to even describe it. At a more practical level, Nobel Prize–winner Francis Crick and his collaborator Christof Koch tried to define the neuronal basis of consciousness; their efforts are reviewed briefly in the following article: C. Koch, "What Is Consciousness?" *Nature* 557 (2018): S9–S12.
2. H. Watanabe, T. Fujisawa, and T. W. Holstein, "Cnidarians and the Evolutionary Origin of the Nervous System," *Development, Growth, and Differentiation* 51, no. 3 (2009): 167–183.
3. C. Dupre and R. Yuste, "Non-overlapping Neural Networks in Hydra vulgaris," *Current Biology* 27 (2017): 1085–1097.

2. ORGANIZATION OF THE HUMAN NERVOUS SYSTEM: FROM NERVES TO NEURONS

1. Nevertheless, it is interesting to note that philosophers in ancient Greece recognized the general function of the brain, although their

ideas were not widely accepted. Alcmaeon of Croton wrote several tracts from 500 to 450 BCE on physiology and psychology and is believed to be the first to identify the brain as the seat of understanding and to distinguish understanding from perception. An excellent treatise on Alcmaeon and his many accomplishments can be found in the Stanford Encyclopedia of Philosophy, which was updated on December 27, 2018. Herophilos (335–280 BCE) dissected the brain and described the course of the nerves that emerged from the brain and spinal cord. Considered the father of anatomy, he concluded from his studies that the brain was the central organ of sensation. He wrote many treatises on anatomy that unfortunately were lost when the library at Alexandria was destroyed.

2. For more information on the composition and organization of the nervous system, readers can consult one of the many excellent textbooks on human anatomy. *Clinically Oriented Anatomy*, by K. L. Moore and A. F. Dalley (Baltimore, MD: Lippincott Williams and Wilkins, 2017), contains a lot of information on the nervous system. There are also many excellent diagrams and illustrations available online.
3. We are following the convention that there are twelve cranial nerves, but this is not correct because nerve XI, the spinal accessory, innervates two muscles in the neck via motor neurons whose cell bodies are located in the cervical region of the spinal cord. Consequently, it functions as a spinal nerve.
4. Each spinal nerve provides the primary innervation to a dermatome, but there is some overlap due to small branches from the spinal nerve above and below.
5. A major figure in understanding this relationship was René Descartes, who in his 1664 *Treatise of Man* postulated that nerves provided a direct connection between external events and the brain. This idea was revolutionary because it showed that pain was a manifestation of activities within the nervous system. The focus on relieving pain could now be placed on preventing the transmission of pain signals along nerves.
6. We use the term "information" simply to mean signals that will ultimately elicit an outcome. In most cases, the signals are the electrical action potentials. We will learn much more about action potentials in subsequent chapters.

7. Santiago Ramon y Cajal was awarded the Nobel Prize in Physiology and Medicine in 1906; information about his many accomplishments can be found by searching his name online.
8. Journal articles can be difficult to read due to the density and volume of information, but these two good sources will help readers understand the role of nociceptive neurons: A. E. Dubin and A. Patapoutian, "Nociceptors: The Sensors of the Pain Pathway," *Journal of Clinical Investigation* 120 (2010): 3760–3772; P. J. Albrecht and F. L. Rice, "Role of Small-Fiber Afferents in Pain Mechanisms with Implications on Diagnosis and Treatment," *Current Pain and Headache Reports* 14 (2010): 179–188.

3. PAIN: PERCEPTION AND ATTRIBUTION

1. K. Hwang et al., "The Human Thalamus Is an Integrative Hub for Functional Brain Networks," *Journal of Neuroscience* 37 (2017): 5594–5607.
2. We are using a third-order neuron as a convenient construct to represent what is actually a complex circuit. At this point, we will also consider perception to be an awareness of a sensation, be it touch, sound, or pain, but acknowledge that this is not precise. Perception is indeed an awareness, but it is not necessarily associated with experiencing a sensation. This distinction between awareness and experience is especially important for understanding pain and will be discussed later in the book.
3. To learn more about these seminal studies, read W. Penfield and T. Rasmussen, *The Cerebral Cortex of Man* (New York: Macmillan, 1950).
4. Interestingly, stimulation of the postcentral gyrus does not elicit pain, affirming that its role is to identify the location of the lesion. Another remarkable finding by Penfield and Rasmussen that is discussed in note 2 is that stimulating the precentral gyrus in patients caused movements of the limbs and face on the opposite side. When these responses were mapped along the gyrus, they formed a motor homunculus in which the hand and face are exaggerated relative to the rest of the body. This explained why a stroke victim is paralyzed on the side opposite to the stroke. It also indicated that we can willfully move a limb or finger, or make a face, because the brain selectively activates the (upper) motor neurons in that region of the homunculus. The axons of these neurons

descend, ultimately resulting in the activation of (lower) motor neurons in the brain stem or ventral horn of the spinal cord. Action potentials from these neurons course through peripheral nerves to elicit contraction of the muscles to achieve the selected action.

5. The immediate response will occur via the first-order A-delta neurons and this will result in acute pain and the initial withdrawal reflex. Because the pathway to the motor neuron is shorter than that to the brain, the withdrawal actually occurs *before* the pain is perceived and this arrangement works well since protecting the finger from additional damage should take precedence. The acute pain is immediately superseded by pain that is mediated via the peripheral and central processes of the C-type nociceptive neurons. These neurons are involved in prolonging the pain and in maintaining the withdrawal in response to additional stimuli to the injury site. A complication is the idea that the A-delta fibers block the inputs for pain that is only overcome by the activation of the C-type nociceptive neurons. The concept was originally proposed by R. Melzack and P. Wall, "Pain Mechanisms: A New Theory," *Science* 150 (1965): 971–979 as the gate-control theory of pain. This theory has been revised because much evidence has shown that many factors regulate the passage of pain impulses to the brain; these factors will be discussed in chapter 8.

4. THE MOLECULAR NEUROBIOLOGY OF PAIN

1. There are differences between the cell bodies of sensory neurons, but their functions are determined by what occurs at their terminals. Thus, the neurons that respond to deep touch, light touch, stretch, positron sense, etc. are each associated with a complex structure that envelops the terminal and converts a specific stimulus into an event that initiates an action potential. Our ability to "sense" our environment is governed by the relatively narrow range to which each receptor responds. The terminals of the nociceptive neurons, and perhaps those involved in itch, are directly exposed to their surroundings and are therefore uniquely situated to respond to a lesion that damages cells in the surrounding tissues.
2. An ion is an atom or group of atoms that carries a net positive (+) or negative (-) electric charge that is denoted by a superscript. Some ions,

such as calcium, have two positive charges (++). This differs from the concept of polarity, in which one end of a molecule is more positive or negative due to an unequal sharing of electrons.

3. The activation of a kinase (and many other enzymes) is tightly controlled because the active kinase will catalyze reactions that can radically alter the properties of the cell. Consequently, the active site of the kinase, i.e., that sequence of amino acids that catalyzes a reaction, is not accessible because it is folded within the three-dimensional conformation of the protein. Exposed on the surface of the kinase is a binding site for a small, specific ligand. When the ligand binds to the site, it induces a conformational change such that the protein unfolds exposing the active region and the onset of the catalytic activity.
4. This description of an action potential is sufficient to understand how electrical signals communicate information from the site of a lesion to the brain. For a more detailed explanation of the molecular processes underlying action potentials and how these are measured, readers can consult any physiology textbook. There are also very good videos online that show how an action potential is generated.
5. Sodium channels are extremely important in information signaling in the nervous system. For an overview on their activation and function, see T. Scheuer, "Regulation of Sodium Channel Activity by Phosphorylation," *Seminars in Cell and Developmental Biology* 22, no. 2 (2011): 160–165.
6. The speed at which an action potential propagates along an axon is determined by the extent to which the axon is myelinated. Myelin is a complex mixture of lipids and proteins that is synthetized by the glial cells that surround and protect axons. The greater the amount of myelin, the greater the speed of propagation. Motor axons are heavily myelinated, whereas the axons of nociceptive neurons are lightly myelinated. Myelin is important in disease states. Readers should consult this excellent review: K. Susuki, "Myelin: A Specialized Membrane for Cell Communication," *Nature Education* 3 (2010): 59.
7. Some neurons communicate via electrical synapses in which an action potential directly activates multiple follower cells without interruption. Electrical synapses are not regulated and they allow a single neuron to activate a large number of follower cells resulting in a global effect. This is useful for the release of hormones or other agents from a gland.

Chemical synapses provide a much more focused and discrete means of communication and are tightly controlled.

8. For an important perspective on the complex roles of sodium channels, see S. R. Levinson, S. J. Luo, and M. A. Henry, "The Role of Sodium Channels in Chronic Pain," *Muscle & Nerve* 46 (2012): 155–165.
9. Tetrodotoxin is an extremely potent neurotoxin found in the liver of puffer fish. The meat of the puffer fish is a delicacy in Japan but must be specially prepared to avoid the poison. The potential danger probably enhances the experience.

5. ADAPTATION

1. Neurons in the central nervous system have a very limited capacity to reproduce, but they can undergo a reorganization. For example, when a finger is amputated, the connections to that finger in the homunculus are somehow redirected to the adjacent fingers making them more receptive. Likewise, a loss of sight typically results in increased acuity in hearing.
2. Small peptides are not synthesized in cells, but their sequence is contained within much larger proteins. When the peptide is needed, enzymes clip the peptide sequence from the protein. In some cases, several peptide sequences are present in a single protein.
3. For a comprehensive study on the roles of bradykinin in pain signaling, see K. J. Paterson et al., "Characterisation and Mechanisms of Bradykinin-evoked Pain in Man Using Iontophoresis," *Pain* 154 (2013): 782–792.
4. NGF is a highly conserved peptide that was first isolated by Nobel Laureates Rita Levi-Montalcini and Stanley Cohen in 1956. The story behind its discovery makes interesting reading and is summarized in the following article: R. Levi-Montalcini and P. U. Angeletti, "Nerve Growth Factor," *Physiological Reviews* 48 (1968): 439–565. NGF is essential for the growth of sensory and sympathetic neurons during development but has additional functions in adults. It contributes to the initiation of pain in response to an inflammation by binding to its tropomyosin receptor kinase A (or TrkA) located on the terminals of nociceptive neurons. NGF also has a very important role in perpetuating pain by altering the synthesis of proteins in the cell body.

5. T. Rosenbaum and S. A. Simon, "TRPV1 Receptors and Signal Transduction," in *TRP Ion Channel Function in Sensory Transduction and Cellular Signaling Cascades*, ed. W. B. Liedtke and S. Heller (Boca Raton, FL: CRC Press/Taylor and Francis, 2007). See also D. Julius, "TRP Channels and Pain," *Annual Review of Cell and Developmental Biology* 29 (2013): 355–384.
6. Eating hot chili peppers can induce a "high" in response to the initial intense pain. This response to a particular type of pepper diminishes, so people who enjoy this feeling are always searching for hotter peppers.
7. The regulation of NMDA channels is unique in several respects, and they have a significant role in many aspects of pain. For a review, read M. L. Blanke and A. M. J. VanDongen, chapter 13 in *Activation Mechanisms of the NMDA Receptor*. NCBI Bookshelf. A service of the National Library of Medicine, National Institutes of Health.
8. Many of the kinases and other factors involved in the posttranslational and phenotypic changes have been identified by Clifford Woolf and his colleagues. For a more comprehensive description of these events see A. Latremoliere and C. J. Woolf, "Central Sensitization: A Generator of Pain Hypersensitivity by Central Neural Plasticity," *Journal of Pain* 10 (2009): 895–926.

6. MOLECULAR SIGNALS FOR PERSISTENT PAIN

1. For the rationale behind using *Aplysia* sensory neurons to study pain, see R. T. Ambron and E. T. Walters, "Priming Events and Retrograde Injury Signals," *Molecular Neurobiology* 13 (1996): 61–96.
2. Y-J Sung, E. T. Walters, and R. T. Ambron, "A Neuronal Isoform of Protein Kinase G Couples Mitogen-Activated Protein Kinase Nuclear Import to Axotomy-Induced Long-Term Hyperexcitability in Aplysia Sensory Neurons," *Journal of Neuroscience* 24 (2004): 7583–7595.
3. Y-J Sung, D. T. W. Chiu, and R. T. Ambron, "Activation and Retrograde Transport of PKG in Rat Nociceptive Neurons After Nerve Injury and Inflammation," *Journal of Neuroscience* 141 (2006): 697–709. See also C. Luo et al., "Presynaptically Localized Cyclic GMP-dependent Protein Kinase 1 Is a Key Determinant of Spinal Synaptic Potentiation and Pain Hypersensitivity," *PLoS Biology* 10 (2012): e1001283.

4. C. Aloe et al., "Nerve Growth Factor: From the Early Discoveries to Its Potential Clinical Use," *Journal of Translational Medicine* 10 (2012): 239–254.

7. THE SOURCES OF PAIN

1. For a more thorough description of the work with mirrors, see the excellent book by V. S. Ramachandran and S. Blakeslee, *Phantoms in the Brain: Probing the Mysteries of the Human Mind* (New York: HarperCollins, 1998).
2. For a comprehensive review, see J. Zhang and J. An, "Cytokines: Inflammation and Pain," *International Journal of Clinical Anesthesiology* 45 (2007): 27–37.
3. How the cytokines and perhaps other agents elicit pain at points along the axon is not known. In particular, it is not clear how their receptors get into the membrane of the axon. They could be inserted into the axonal membrane in the cell body and then migrate slowly along the plane of the membrane. Alternatively, we know that vesicles containing the receptors are being transported rapidly within the axon to the terminals. Some of these could be diverted to fuse with the axonal membrane at the site of the inflammation. Determining which mechanism is correct has implications for treating these types of inflammatory pain.
4. Q. Xu and T. L. Yaksh, "A Brief Comparison of the Pathophysiology of Inflammatory Versus Neuropathic Pain," *Current Opinion in Anesthesiology* 24 (2011): 400–407.
5. An early inference from the homunculus was that pain from somatic structures can be directly sensed, whereas pain from visceral structures cannot. This viewpoint has been amended because there is a small representation of the back of the oral cavity that is composed of structures that originate from both somatic and visceral sources and where pain can be sensed.
6. The interaction between the afferent visceral neurons and the motor neurons of the autonomic nervous system was thought to maintain homeostasis, which is postulated to be the optimal resting state of the body. This is an oversimplification. These interactions did not develop to maintain a single state of homeostasis but rather as a way of constantly adjusting the functions of the viscera to accommodate the conditions

present at each moment. This more dynamic view is more in keeping with meeting the ever-changing events in our lives that require constant adjustments to our heart rate, blood pressure, etc.

7. This explanation is reasonable but simplistic because there is evidence that other groups of neurons in the spinal cord are involved in processing nociceptive information from the viscera. For example, see V. Krolov et al., "Functional Characterization of Lamina X Neurons in Ex Vivo Spinal Cord Preparation," *Frontiers in Cellular Neuroscience* 11 (2017): 342–352.
8. Not every spinal nerve has branches of the visceral nerves. The visceral branches enter the spinal cord only within nerves at levels Thoracic-1 to about Lumbar-3.

8. THE EXTERNAL MODULATION OF PAIN: DESCENDING SYSTEMS

1. For a more detailed description of this fascinating story, see J. Goldberg, *Anatomy of a Scientific Discovery: The Race to Find the Body's Own Morphine* (New York: Skyhorse, 2013).
2. Cells and tissue contain hundreds of enzymes that degrade damaged proteins and other cell constituents. These enzymes are normally enclosed in membranous compartments within cells, but they are released during the homogenization process. That the endorphin survived was fortunate indeed.
3. The word *opioid* is reserved for the compounds in the brain. An *opiate* is an exogenous compound, such as morphine, that mimics the function of the endogenous compounds. Neurons containing opioids are called *opioidergic*. Naloxone binds to the receptor with such high affinity that it will displace a bound opioid or opiate. This has been put to use clinically and a nasal spray containing naloxone can rapidly relieve opiate-induced respiratory distress after an overdose.
4. M. Brownstein, "Review: A Brief History of Opiates, Opioid Peptides, and Opioid Receptors," *Proceedings of the National Academy of Sciences of the United States of America* 90 (1993): 5391–5393.
5. The periaqueductal gray is named after its relationship to s series of reservoirs called ventricle that lie deep in the brain. There are 4 ventricles and the 3rd and 4th are interconnected by a narrow tube known as

the aqueduct. the neurons adjacent to this duct are logically called peri (next to) aqueductal.

6. G. W. Pasternak and Y. X. Pan, "Mu Opioids and Their Receptors: Evolution of a Concept," *Pharmacology Review* 65 (2013): 1257–1317.
7. For a good general review of the subject, see E. Roberts, "Basic Neurophysiology of GABA," *Scholarpedia* 2 (2007): 3356.
8. For a comprehensive analysis of GABA receptors, see E. Sigel and M. E. Steinmann, "Structure, Function, and Modulation of GABAA Receptors," *International Journal of Biological Chemistry* 287 (2012): 40224–40231.
9. Lyrica is a calcium channel antagonist that also increases the activity of the enzyme that synthesizes GABA, thereby elevating GABA levels. Drugs that promote activity are called *agonists*, as opposed to *antagonists*, which block activity. Gabapentin, despite its name, does not modulate GABA levels. Rather, it blocks a subset of calcium channels and is sometimes prescribed as an adjunct treatment for pain.
10. H. Obata, "Analgesic Mechanisms of Antidepressants for Neuropathic Pain," *International Journal of Molecular Sciences* 18 (2018): 2483–2495.

9. ALLEVIATING PAIN: THE PHARMACOLOGICAL APPROACH

1. A formulation of opium called laudanum was devised by Theophrastus Phillippus Aureolus Bombastus von Hohenheim (aka Paracelsus, 1493–1541). Laudanum was lauded because it relieved pain without diminishing function and many writers and artists of the Victorian period used it. For example, Samuel Taylor Coleridge's most famous poem, "Kubla Khan," was written after an intense laudanum-induced dream and the poet Elizabeth Barrett Browning depended on laudanum for inspiration. However, it did not go unnoticed that opium could have deleterious side effects, including addiction. The problem was exacerbated as recipes for laudanum evolved over time and some included various amounts of alcohol. The latter made laudanum attractive for recreational use and it remained a popular remedy into the twentieth century.
2. Not every derivative was beneficial. In 1874, Charles Adler Wright synthesized diacethyl-morphine, which we know as heroin. Heroin was more effective as a cough suppressor than an analgesic, and it became

widely available. It was also much more addictive than morphine. Drugs that had been used for centuries to relieve pain were now used recreationally to get "high" or were overprescribed for pain.

3. The resin secreted from the flowers of the female plants is known as hashish, which is a more potent psychoactive agent than marijuana.
4. See E. J. Rahn and A. G. Hohmann, "Cannabinoids as Pharmacotherapies for Neuropathic Pain: From the Bench to the Bedside," *Journal of the American Society for Experimental Neurotherapeutics* 6 (2009): 713–737.
5. The name anandamide is taken from the Sanskrit word *ananda*, which means "joy, bliss, delight," which is apt given its effects on various centers in the brain.
6. Rahn and Hohmann, "Cannabinoids as Pharmacotherapies for Neuropathic Pain," 713–737. Research on the cannabinoid system is progressing rapidly and many new results will be forthcoming. For a good general review, see S. Vuckovic et al., "Cannabinoids and Pain: New Insights from Old Molecules," *Frontiers in Pharmacology* 9 (2018): 1259.
7. For a thorough discussion of the CB2-receptor system, see C. Turcotte et al., "The CB2 Receptor and Its Role as a Regulator of Inflammation," *Cellular and Molecular Life Sciences* 73 (2016): 4449–4470.
8. Updates on cannabidiol can be obtained online from the National Center for Biotechnology Information.
9. For a nice description of this barrier, see M. Blanchette and R. Daneman, "Formation and Maintenance of the BBB," *Mechanisms of Development* 138 (2015): 8–16.
10. D. Jamero et al., "The Emerging Role of NMDA Antagonists in Pain Management," *U.S. Pharmacist* 36 (2011): HS4–HS8.
11. These are referred to as ADME properties for drug development. Many online references describe each in detail.
12. There are significant differences in the types of trials that are required before a given drug can be marketed. Analgesic drugs face greater hurdles than do drugs for cancer, which can sometimes enter what is known as a fast track. Drugs that take this route are not tested as rigorously and are therefore much less expensive to produce. Bottom line: it is more profitable for a drug company to develop a drug for cancer rather than an analgesic. Details about all of the requirements can be found in online publications from the FDA.

10. THE NEUROMATRIX

1. For one of the earliest references describing this fascinating phenomenon, see J. I. Rubbins and E. D. Friedman, "Asymbolia for Pain," *Archives of Neurology and Psychiatry* 60 (1948): 554–573.
2. Melzac and Casey were among the first to propose that the emotional determinants of pain were properties that emerged from functions in the brain and that they were distinctly different from those responsible for the sensory and discriminative dimensions of pain. See R. Melzack and K. L. Casey, "Sensory, Motivational, and Central Control Determinants of Pain," in *The Skin Senses*, ed. D. Kenshalo (Springfield: Thomas, 1968), 423–439.
3. See Melzack and Casey, "Sensory, Motivational, and Central Control Determinants of Pain," 423–439.
4. Affect is a generic term for the expression of emotions to others via facial features, tone of voice, etc. In this case, the affect is not for the world outside, but is the imposition of mood, anxiety, fear, etc. on the expression of pain.
5. Melzack, R. "Phantom Limbs and the Concept of a Neuromatrix," *Trends in Neurocience* 13 (1990): 88–92.
6. D. L. Morton, J. S. Sandhu, and A. K. P. Jones, "Brain Imaging of Pain: State of the Art," *Journal of Pain Research* 9 (2016): 613–624.
7. There is evidence that that there are two outputs from the thalamus; one to the sensory cortex for the localization of the lesion, and another to the affective areas. See B. Kulkarni, "Attention to Pain Localization and Unpleasantness Discriminates the Functions of the Medial and Lateral Pain Systems," *European Journal of Neuroscience* 21 (2005): 3133–3142.
8. A lesion to the amygdalae results in Kluver-Bucy syndrome that was named after Heinrich Kluver and Paul Bucy who characterized the resultant changes in behavior. One interesting finding is that there must be connections between the olfactory system and the amygdala because fear was elicited when a particular odor had been associated with a traumatic event.
9. *A nucleus* in the central nervous system is a defined group of neuronal cell bodies whose axons project to other areas of the brain and whose dendrites receive information from other areas. These neurons will

typically contribute to a single function. A ganglion is its counterpart in the peripheral nervous system.

10. For another perspective, see E. Brodin, M. Ernberg and L. Olgart, *Neurobiology: General Considerations—From Acute to Chronic Pain," Den Norske Tannlegeforenings Tidende* 126: 28–33.
11. D. Biro, "Is There Such a Thing as Psychological Pain? and Why It Matters," *Culture, Medicine, and Psychiatry* 34 (2010): 658–667.
12. We could speculate that *all* painfulness requires the activation of neurons in the thalamus. This would be consistent with the theory that a subgroup of thalamic neurons contributes directly to the affective modulation of pain.

11. THE BRAIN AND PAIN

1. Distinguishing nonsuicidal self-injury (NSSI) from injury caused by attempting suicide can be difficult. Criteria for NSSI include five or more days of self-inflicted harm over the course of one year without suicidal intent. The motivation must be an effort to obtain temporary relief from intense negative feelings caused by a sense of failure, rejection, or self-loathing.
2. Self-harm is practiced in Hinduism by the ascetics and in Catholicism and Islam where it is called mortification of the flesh and is used as a form of atonement for sin or misdeeds.
3. Note that the imaging techniques have sufficient resolution to recognize areas that are smaller than the areas delineated by Brodmann.
4. L. Q. Uddin et al., "Structure and Function of the Human Insula," *Journal of Clinical Neurophysiology* 34 (2017): 300–306.
5. There is much that is not known about the functions of the insular cortex. What is fascinating is that neurons in the IC are also activated by observing others in pain, which is an expression of empathy.
6. A meta-analysis is a statistical evaluation of data collected from several studies published by others. By collecting data from multiple sources, the number of subjects is increased and so is the validity of the findings. With this in mind, see P. Yuana and N. Raz, "Prefrontal Cortex and Executive Functions in Healthy Adults: A Meta-Analysis of Structural Neuroimaging Studies," *Neuroscience & Biobehavioral Reviews* 42 (2014): 180–192.

7. S. Kamping, "Contextual Modulation of Pain in Masochists: Involvement of the Parietal Operculum and Insula," *Journal of Pain* 157 (2016): 445–455.
8. The situation is more complicated because the masochist group also had increased activity in regions of the brain that are involved in processing visual information. This was expected since the attenuation of pain depended on seeing the pictures. Regions involved in memory were also activated.
9. We have already discussed the fact that pain can be relieved by a treatment that has no actual effect on pain; this complicates efforts to identify drugs and other treatments that appear to have analgesic effects. Only carefully controlled studies and statistical analyses of the results can distinguish the real from the sham.
10. There are many papers on the placebo effect; the following article gives a good review and provides references for further reading: T. D. Wager and L. Y. Atlas, "The Neuroscience of Placebo Effects: Connecting Context, Learning and Health," *Nature Reviews Neuroscience* 16 (2015): 403–418.
11. Wager and Atlas, "The Neuroscience of Placebo Effects," 403–418.
12. The words "hypnosis" and "hypnotism" both derive from the term "neurohypnotism" (nervous sleep) that was coined by Étienne Félix d'Henin de Cuvillers in 1820. Hypnotism was popularized by the Scottish surgeon James Braid, who used it in his practice and promoted what he saw as its biological and physical benefits. His recognizing the connections between hypnotism and the Oriental practices of meditation and yoga were prescient, and we will discuss these in detail in chapters dealing with the importance of meditation in relieving pain. There are many very interesting online sources that discuss Mesmer, Braid, and the history of hypnosis.
13. H. Jiang et al., "Brain Activity and Functional Connectivity Associated with Hypnosis," *Cerebral Cortex* 27 (2017): 4083–4093.
14. Alternative practices for pain relief were used to supplement the narcotic and supernatural approaches and one Eastern import was acupuncture. Willem ten Rhijne was especially interested in this unique therapeutic technique; in 1683, he published an extensive description of acupuncture that he learned from his Japanese mentors. According to his writing, acupuncture depended on the precise insertion of fine

needles to influence "meridians" that coursed through the body. To date, there is no anatomical structure that corresponds to the meridians.

15. Han, J-S, "Acupuncture and Endorphins," *Neuroscience Letters* 361 (2004): 258–261.

12. THE MIND REGULATING THE MIND

1. R. Staud, "Brain Imaging in Fibromyalgia Syndrome," *Clinical and Experimental Rheumatology* 29, suppl. 69 (2011): S109–S117. See also D. L. Morton, S. Sandhu, and A. K. P. Jones, "Brain Imaging of Pain: State of the Art," *Journal of Pain Research* 9 (2016): 613–624.
2. Freud updated his theory to postulate the roles of the id, ego, and superego in controlling behavior. Readers interested in learning more about Freud and his views regarding pain will find many excellent publications online.
3. J. E. Sarno, *The Mindbody Prescription: Healing the Body, Healing the Pain* (New York: Grand Central Publishing, 1999). This influential book offers the rationale that chronic pain is due primarily to suppressed emotions. Dr. Sarno, an expert on lower back pain, postulated that chronic pain is due to a tension myositis syndrome in which the conscious mind acting via the autonomic nervous system elicits localized minor constriction of arterioles that induces ischemia (reduced blood supply) and pain. However, vasoconstriction happens all the time without pain and there is no evidence to support the implication that the brain has such fine control over the autonomic nervous system that it can cause a localized constriction in vessels within specific organs or structures.
4. E. Scarry, *The Body in Pain: The Making and Unmaking of the World* (Oxford: Oxford University Press, 1987). For a more clinical study, see C. Oliviola and E. Shafir, "The Martyrdom Effect: When Pain and Effort Increase Prosocial Contributions," *Journal of Behavioral Decision Making* 26 (2013): 91–105.
5. See articles on martyrdom by L. Stephanie Cobb, who teaches New Testament and early Christianity studies at the University of Richmond.
6. A. Lazaridou et al., "The Impact of Anxiety and Catastrophizing on Interleukin-6 Responses to Acute Painful Stress," *Journal of Pain Research* 11 (2018): 637–647. See also J. A. Sturgeon, "Psychological

Therapies for the Management of Chronic Pain," *Psychology Research and Behavior Management* 7 (2014): 115–124.

7. J. C. Felger, "Imaging the Role of Inflammation in Mood and Anxiety-related Disorders," *Current Neuropharmacology* 16 (2018): 533–558.
8. J. L. Bantick et al., "Imaging How Attention Modulates Pain in Humans Using Functional MRI," *Brain* 128 (2002): 310–319.
9. L. M. McCracken and K. E. Vowles, "Acceptance and Commitment Therapy and Mindfulness for Chronic Pain," *American Psychologist* 69 (2014): 178–187.
10. L. A. Slagter, H. A. Dunne, and R. J. Davidson, "Attention Regulation and Monitoring in Meditation," *Trends in Cognitive Science* 12 (2008): 163–169.
11. M. Boccia, L. Piccardi, and P. Guariglia. "The Meditative Mind: A Comprehensive Meta-Analysis of MRI Studies," *BioMed Research International* (2015). Article ID 419808.
12. J. Kabat-Zinn, "An Outpatient Program in Behavior Medicine for Chronic Pain Patients Based on the Practice of Mindfulness Meditation," *General Hospital Psychiatry* 4 (1982): 33–47.
13. Pierre Paul Broca, a French surgeon and anatomist, studied patients in the 1850s who had lesions in the left cerebral cortex just in front of the face in the motor homunculus. These patients were unable to speak, although the vocal apparatus was intact. The condition was called Broca's aphasia; we now know it is caused by damage to the cortical neurons. This was the first anatomical proof of localization of brain function. Moreover, the lesion affected speech only if it was located on the left side, indicating that the right and left hemispheres have different functions.
14. Many books describe these disciplines and practices in detail; we are only presenting a general outline.
15. L. F. Haas, "Hans Berger (1873–1941), Richard Caton (1842–1926), and Electroencephalography," *Journal of Neurology, Neurosurgery & Psychiatry* 74 (2003): 9.
16. F. Zeidan et al., "Brain Mechanisms Supporting the Modulation of Pain by Mindfulness Meditation," *Journal of Neuroscience* 31, no. 14 (2011): 5540–5548.
17. F. Zeidan et al., "Mindfulness-Meditation-Based Pain Relief Is Not Mediated by Endogenous Opioids," *Journal of Neuroscience* 36 (2016): 3391–3397.

18. R. C. deCharms et al., "Control over Brain Activation and Pain Learned by Using Real-time Functional MRI," *Proceedings of the National Academy of Sciences of the United States of America* 102 (2005): 18628–18631.

13. PAIN MANAGEMENT: PRESENT AND FUTURE

1. S. R. Levinson, S. Luo, and M. A. Henry, "The Role of Sodium Channels in Chronic Pain," *Muscle & Nerve* 46 (2012): 155–165.
2. For a discussion of targets for the development of analgesics, we recommend A. S. Yekkirala et al., "Breaking Barriers to Novel Analgesic Drug Development," *Nature Reviews Drug Discovery* 16 (2017): 545–564.
3. Y.-J. Sung and R. T. Ambron, "Pathways That Elicit Long-Term Changes in Gene Expression in Nociceptive Neurons Following Nerve Injury: Contributions to Neuropathic Pain," *Neurological Research* 26 (2004): 195–203.
4. This was accomplished as an extraordinary collaboration with the Department of Medicine at Columbia University and Ramy Farad and Jeremy Greenwood at Schrodinger Inc. New York. Y.-J. Sung et al., "A Novel Inhibitor of Active Protein Kinase G Attenuates Chronic Inflammatory and Osteoarthritic Pain," *Pain* 158 (2020): 822–832.
5. L. Aloe et al., "Nerve Growth Factor: From the Early Discoveries to the Potential Clinical Use," *Journal of Translational Medicine* 10 (2012): 239–254.
6. In addition to the traditional approach that relies on synthesizing small molecules as analgesics, many drug companies are using monoclonal antibodies that recognize specific regions in the target. A drug whose name ends in "mab" is a monoclonal antibody.
7. M. K. Patel, A. D. Kaye, and R. D. Urman, "Tanezumab: Therapy Targeting Nerve Growth Factor in Pain Pathogenesis," *Journal of Anaesthesiology Clinical Pharmacology* 34 (2018): 111–116.
8. S. M. Farrell, A. Green, and T. Aziz, "The Current State of Deep Brain Stimulation for Chronic Pain and Its Context in Other Forms of Neuromodulation," *Brain Sciences* 8 (2018): 158–177.
9. Dejerine–Roussy, or thalamic pain syndrome, is a condition that develops after a stroke damages neurons in the thalamus. It is characterized by abnormal sensations, such as tingling, and severe allodynia that is very difficult to treat.

10. J. M. R. Delgado, *Physical Control of the Mind: Toward a Psychocivilized Society* (New York: Harper and Row, 1969) and J. M. R. Delgado, "Free Behavior and Brain Stimulation," *International Review of Neurobiology* 6 (1964): 349–449.
11. The text provides a relatively brief discussion of optogenetics as a potential method of controlling pain. For a much more comprehensive view, read A. Guru et al., "Making Sense of Optogenetics," *International Journal of Neuropsychopharmacology* 1–8 (2015).
12. S. G. Boccard, et al., "Deep Brain Stimulation of the Anterior Cingulate Cortex: Targeting the Affective Component of Chronic Pain," *Neuroreport* 25, no. 2 (2014): 83–88.

INDEX

Printed and bound by CPI Group (UK) Ltd, Croydon, CR0 4YY

29/04/2024

14490874-0001